Human-Centered Design for Health Care Safety

Human-Centered Design for Health Care Safety describes a conceptual framework (SEIPS or Systems Engineering Initiative for Patient Safety) based on the Human Factors and Ergonomics (HFE) discipline and associated human-centered design (HCD) approaches and methods that can help to create innovative solutions for enhancing health care safety. Health care safety remains a major concern around the world. With global crises, such as economic downturns and the COVID-19 pandemic, we have observed deterioration in many aspects of patient safety, as well as major problems for health care workers, including burnout, stress and issues of occupational safety and health. Therefore, we need renewed attention on how to design health care work systems and care processes to improve safety for patients and health care workers.

This book describes key aspects of a widely adopted HFE systems approach to health care safety and methods for HCD in health care. It provides insights into challenges for future research in health care safety. The benefits of SEIPS are promoted throughout the text and reveal how they can lead to improvement in the design of technologies, environments and the rest of the work system; thereby, supporting, instead of hindering, the activities of all care team members, including patients and care partners. Case studies reveal examples of SEIPS in action and various domains of health care safety. The book provides practical tips on how SEIPS is implemented in practice. The readers will benefit from this book as it describes the foundation for SEIPS and HCD and multiple applications across varied settings and domains of health care safety. They will receive a toolkit for enacting solutions in their work environment.

This exciting and topical title will appeal to students and professionals in the fields of health care, medicine, nursing, pharmacy, other health professions and pharmaceuticals. It will also appeal to industrial and systems engineers, biomedical engineers, HFE professionals and designers.

Human Factors and Ergonomics
Series Editors
Waldemar Karwowski, University of Central Florida, Orlando, USA
Pamela McCauley, University of Central Florida, Orlando, USA

PUBLISHED TITLES

Human Factors in System Design, Development, and Testing
D.Meister and T. Enderwick

Introduction to Human Factors and Ergonomics for Engineers, Second Edition
M. R. Lehto

Macroergonomics: Theory, Methods and Applications
H.Hendrick and B. Kleiner

Modeling Sociocultural Influences on Decision Making: Understanding Conflict, Enabling Stability
Joseph V. Cohn, Sae Schatz, Hannah Freeman, and David J. Y. Combs

Practical Speech User Interface Design
James R. Lewis

The Science of Footwear
R. S. Goonetilleke

Skill Training in Multimodal Virtual Environments
M.Bergamsco, B. Bardy, and D. Gopher

Smart Clothing: Technology and Applications
Gilsoo Cho

Theories and Practice in Interaction Design
S.Bagnara and G. Crampton-Smith

The Universal Access Handbook
C.Stephanidis

Usability and Internationalization of Information Technology
N. Aykin

User Interfaces for All: Concepts, Methods, and Tools
C. Stephanidis

Variability in Human Performance
T. Smith, R. Henning, and M. Wade

Handbook of Standards and Guidelines in Human Factors and Ergonomics, Second Edition
Waldemar Karwowski, Anna Szopa, & Marcelo M. Soares

Human-Centered Design for Health Care Safety: SEIPS in Action
Pascale Carayon

Human-Centered Design for Health Care Safety

SEIPS in Action

Pascale Carayon

CRC Press
Taylor & Francis Group
Boca Raton London New York

CRC Press is an imprint of the
Taylor & Francis Group, an **informa** business

Designed cover image: Shutterstock

First edition published 2026
by CRC Press
2385 NW Executive Center Drive, Suite 320, Boca Raton FL 33431

and by CRC Press
4 Park Square, Milton Park, Abingdon, Oxon, OX14 4RN

CRC Press is an imprint of Taylor & Francis Group, LLC

ISBN: 978-1-032-49864-5 (hbk)
ISBN: 978-1-032-76897-7 (pbk)
ISBN: 978-1-003-48031-0 (ebk)

DOI: 10.1201/9781003480310

Typeset in Times
by SPi Technologies India Pvt Ltd (Straive)

Dedication

To my parents, Robert and Michèle Carayon
To my life partner, Peter Hoonakker

Contents

Figures

Tables

About the author

Pascale Carayon, PhD, is Professor Emerita in the Department of Industrial and Systems Engineering at the University of Wisconsin-Madison, United States. She is the Founding Director of the Wisconsin Institute for Healthcare Systems Engineering; she led the interdisciplinary Systems Engineering Initiative for Patient Safety (SEIPS) program at the University of Wisconsin-Madison, United States. She received her Engineer diploma from the Ecole Centrale de Paris, France, and her PhD in Industrial Engineering from the University of Wisconsin-Madison. Dr. Carayon has over three decades of HFE research experience analyzing, designing, and improving complex work systems such as those in health care. In 2024, she was elected to the *National Academy of Engineering* for application of human factors engineering to health care systems to improve patient safety.

Preface

Health care safety remains a major concern around the world. With frequent crises (e.g., economic crises, epidemics and pandemics, regional wars and conflicts), we have observed deterioration in many aspects of patient safety (e.g., delays in diagnosis, health care-associated infections, disparities in medication or vaccine delivery), as well as major problems for health care workers, such as burnout, stress and work-related injuries. Therefore, we need renewed attention on how to design health care work systems and care processes in order to improve safety for both patients and health care workers, i.e. health care safety. This book describes a conceptual framework (SEIPS or Systems Engineering Initiative for Patient Safety) and associated human-centered design (HCD) approaches and methods that can help to create innovative solutions for continuously enhancing health care safety.

Health care safety problems are intrinsically related to the way care is organized, i.e. the design of the work systems and care processes. A systems approach embedded in the Human Factors and Ergonomics (HFE) discipline, such as SEIPS, can help to redesign technologies, environments, work organization and care processes in order to support, instead of hinder, the individual and collective activities of all care team members, e.g. physicians, nurses, pharmacists, patients, care partners and many other stakeholders. A family of SEIPS models has been developed over the years and includes three core components:

1. Work system model with the following interacting elements: *person, tasks, tools and technology, physical environment, the organization and the external environment*. The *person* is at the center of the work system. [Note: in the book, *the work system elements are italicized*.]
2. Multiple outcomes for patients and care partners, health care workers and health care organizations. Health care safety is a combination of patient safety and health care worker safety.
3. Feedback loops between work systems, care processes and outcomes: adaptation, learning and improvement.

Who can benefit from the book? Health care leaders, professionals and safety practitioners need guidance on how to implement a systems approach for improving health care safety. The book provides practical tips, examples and case studies of how SEIPS is implemented in practice. I also discuss challenges in applying SEIPS and HFE for health care safety improvement and cultural and organizational issues of embedding HFE in health care.

Graduate students in health-related disciplines (e.g., health services research, medicine, nursing, pharmacy, physical therapy) and in engineering and design fields with an interest in health and health care (e.g., industrial and systems engineering, biomedical engineering, HFE, industrial design, human–computer interaction) will benefit from this book as it describes the foundation for SEIPS and human-centered design (HCD) and multiple applications in various care settings and safety domains.

Health care researchers and teachers will also benefit from the book as it describes key aspects of a widely adopted systems approach to health care safety (i.e. SEIPS) and methods for HCD in health care. The book can be used as a textbook for courses on patient safety, health care improvement, health care systems engineering and health care HFE. By highlighting difficulties and controversies, the book also provides insights into challenges for future SEIPS application and research in health care safety.

How to read the book? The book brings together a range of knowledge from conceptual and theoretical domains (e.g., systems thinking) to practical approaches (e.g., HCD) and applications (e.g., examples and case studies). Depending on your professional goals and your level of HFE knowledge, you should find something of interest in the book to address health care safety in your projects and various endeavors.

I am a systems thinker; therefore, in writing this book, I created a lot of connections and relationships. Throughout the book, you will find numerous mentions of chapters and sections. The book includes seven chapters that are written separately, but they are all connected to each other.

Chapter 1 describes the broad concept of health care safety, i.e. safety for patients and safety for health care workers. In Chapter 2, I describe systems thinking and emphasize HFE systems approaches. Chapter 3 is dedicated to a deep dive on SEIPS, including the different models of the SEIPS family and its conceptual and theoretical basis in HFE, the Balance Theory, the Sociotechnical Systems (STS) Theory and other models. Chapter 4 is about putting SEIPS in practice through HCD. In Chapter 4, I describe various ways of thinking about HCD and applications of SEIPS in the HCD process, including SEIPS analysis of work systems and care processes and design, implementation and evaluation of work system interventions. Chapter 5 is dedicated to three selected projects with the goal of showing different applications of SEIPS in practice. Chapter 6 includes my reflections on SEIPS and challenges in improving health care safety. I also share ideas for future development of SEIPS. Chapter 7 provides a short conclusion.

SEIPS was developed over many years and will continue to evolve with research and practice. In writing this book, I selected multiple papers, studies and projects that use SEIPS; many of them are related to action-research projects that I have been involved in. I have incorporated a lot of my own experience with SEIPS and health care safety in the book. As applications of SEIPS in research and practice develop and grow, we learn a lot about how to improve health care safety as well as how to use and further develop SEIPS. Therefore, this book is part of a long-term learning journey to produce benefits for all involved in health care safety and for all patients and people around the world. I hope that you will become part of this continuous learning journey.

Acknowledgments

My name is on the front cover of the book. But I wish that publishers would allow a 'cloud' of names to show up on the front cover of books. A book like this is never something done by a single individual; it is the product of work by and interactions between multiple individuals over many years. Of course, this is the case for this book.

First, I want to dedicate this book to my family: my husband (Peter), my mom (Michèle) and my late dad (Robert), my sisters (Laurence and Sophie), my French brother-in-law (Christian), my Dutch sisters-in-law (Loes, Willy and Anneke), my Dutch brother-in-law (Paul), my French nieces (Camille, Clémence, Marie and Clara) and my Dutch nieces and nephews (Mo, Nico, Laura, Anke, Stijn and Jolien). I would not be who I am today without your love, support, patience and help. Peter is my life partner as well as my collaborator in multiple teaching and research projects, including SEIPS. He reviewed multiple drafts of this book; the book was enriched and improved by his feedback and our multiple conversations on SEIPS and health care safety improvement. Thank you, dank u wel, Peter. I also dedicate this book to Missia, Minana and Pierre-Dominique who are no longer with us.

Back in 1985, the late Professor Michael J. (Mike) Smith gave me a chance to do a PhD in Industrial Engineering at the University of Wisconsin-Madison, United States; his decision changed my life. I would like to express my gratitude for everything that Mike has done to help me become and grow as a teacher, a researcher and a better person. This book is dedicated to you, Mike. When you read the book and learn about the work system model and the Balance Theory, you will hopefully understand the influence and impact of Mike on SEIPS.

Thank you to Ann Schoofs Hundt for being such a great friend and collaborator on so many SEIPS projects and research and teaching activities. Our numerous planning sessions, workshops, seminars, field work, organizational development, etc., could not have led to such great work and impact without your amazing contribution and your dedication. Your deep knowledge in HFE, organizational design and technology combined with your health care experience had a huge impact on SEIPS. All of my gratitude for your participation and engagement in our various SEIPS endeavors! More importantly, your friendship means a lot to me.

The projects and research described in this book involved many people and organizations. Thank you to my graduate students who participated in multiple projects and are continuing the development and application of SEIPS in their research, teaching and practice, including Ayse Gurses, Anping Xie, Bashar Alyousef, Nana Khunlertkit, Kerry McGuire, Abby Wooldridge, Sarah Kianfar, Yaqiong Li, Megan Salwei, Betsy Hose and Kathryn Wust. Also, thank you to graduates from the University of Wisconsin-Madison, United States, who have helped to develop and grow SEIPS, in particular Rich Holden, Rupa Valdez, Joy Rivera, Hanna Barton and Rachel Rutkowski. Thank you to my numerous collaborators in health care around the United States, that is physicians, nurses, pharmacists, unit clerks, IT professionals, organizational leaders, patient safety officers, student helpers, patient advocates, etc.

They include Maureen Smith, Shawna Perry, Tosha Wetterneck, Ken Wood, Michelle Kelly, Nasia Safdar, Jim Walker, Manish Shah, Brian Patterson, Mike Pulia, Barbara King, Shannon Dean, Tom Brazelton, Josh Ross, Jackson Musuuza, Scott Springman, Deb Rusy, Ben Eithun, Jonathan Kohler and Paula Dail.

I also want to thank many friends and collaborators. In particular, my friend, Carla Alvarado, played a special role in SEIPS. Her energy and vision helped to develop SEIPS projects and short courses. Her extensive knowledge in HFE and infectious diseases opened many doors to fruitful collaboration and impactful projects. I also want to thank many people that I had a chance to work with and learn from, in particular at the University of Wisconsin-Madison; they include Patti Brennan, Denny Fryback, Nicole Werner, Doug Wiegmann, Jingshan Li, Teresa Thuemling, Dave Gustafson and Ian Robertson. A special 'merci' to Pierre Falzon, in particular for his help in understanding the unique features of the French ergonomics tradition and for our collaboration in international HFE development efforts.

Thank you to those who paved the way early on for research and practice in HFE and health care safety. I want to thank the (late) Sue Bogner; she was very generous in sharing her knowledge and insights on patient safety, such as the early study by Al Chapanis and his colleague on medication safety. A special thank you to David Bates for his international leadership in patient safety and his support along my journey in developing SEIPS and HFE research in health care safety.

In wrapping up the writing of this book, I also want to remember friends and colleagues who are no longer with us. I have learned so much from them and I miss our conversations, in particular with my HFE friends and colleagues: John Wilson from the United Kingdom, Klaus Zink from Germany and Hal Hendrick from the United States. A special thought to my friend and colleague Bentzi Karsh, who passed away too early. I was one of his professors, but soon he became a close collaborator, and I wish that I could have continued to learn a lot more from him. I also miss Bob Wears; his deep and broad knowledge in safety was amazing. What a great friend and colleague who trusted our team early on and participated in our first SEIPS short course!

I have probably forgotten to mention many people; sorry about this! As I release this book to the publisher, I am hopeful that it will contribute to your learning and maybe help you see and approach your work in a different way; if it does, please let me know. I also want to learn from you.

Many people who are interested and involved in health care safety improvement and who read this book are health care workers. Many others are, have been or will be patients. Most are, have been and will be care partners. This book is for all of us.

Abbreviations

ADE	Adverse Drug Event
AHRQ	Agency for Healthcare Research and Quality
APP	Advanced Practice Provider
AVS	After-Visit Summary
BCMA	Bar Code Medication Administration
CDC	Centers for Disease Control and Prevention
CDI	*C. difficile* Infection
CDS	Clinical Decision Support
CMO	Context-Mechanism-Outcome
CPOE	Computerized Provider Order Entry
DCERPS	Developmental Center for Evaluation and Research in Patient Safety
DOD	Department of Defense
ED	Emergency Department
EFM	Electronic Fetal Monitoring
FMEA	Failure Mode and Effects Analysis
FQ	Fluoroquinolones
HAI	Health care-Associated Infection
HCD	Human-Centered Design
HCI	Human–Computer Interaction
HFE	Human Factors and Ergonomics
HFES	Human Factors and Ergonomics Society
HRO	High Reliability Organization
HSIB	Healthcare Safety Investigation Branch
IEA	International Ergonomics Association
ICU	Intensive Care Unit
IOM	Institute of Medicine
ISO	International Organization for Standardization
IT	Information Technology
MEAD	MacroErgonomic Analysis and Design
MRI	Magnetic Resonance Imaging
NASEM	National Academies of Sciences, Engineering, and Medicine
NHS	National Health Service
NIOSH	National Institute for Occupational Safety and Health
NPO	Nothing by mouth
ODAM	Organizational Design and Management
OECD	Organisation for Economic Co-operation and Development
OR	Operating Room
OSHA	Occupational Safety and Health Administration
OTC	Over-The-Counter
PE	Pulmonary Embolism
PICU	Pediatric Intensive Care Unit
PPE	Personal Protective Equipment

PRA	Proactive Risk Assessment
PSIRF	Patient Safety Incident Response Framework
PSLL	Patient Safety Learning Laboratory
QI	Quality Improvement
SD	System Dynamics
SEIPS	Systems Engineering Initiative for Patient Safety
SPD	Sterile Processing Department
SNF	Skilled Nursing Facilities
SPO	Structure-Process-Outcome
SSM	Soft Systems Methodology
STEW	Systems Thinking for Everyday Work
VTE	Venous Thromboembolism
WRMDs	Work-Related Musculoskeletal Disorders

1 Improving Health Care Safety

My approach to improving health care safety is based in the Human Factors and Ergonomics (HFE) discipline, which is defined as 'the scientific discipline concerned with the understanding of interactions among humans and other elements of a system, and the profession that applies theory, principles, data, and methods to design in order to optimize human well-being and overall system performance' (International Ergonomics Association (IEA), 2000). The terms *Human Factors* and *Ergonomics* are used interchangeably; different countries have different traditions and use either the term *Human Factors*, the term *Ergonomics*, or together as a unit (e.g. *HFE*, *HF/E* or *EHF*). As indicated in the International Ergonomics Association (IEA) definition of HFE, it is important to design work systems that 'optimize human well-being and overall system performance' (International Ergonomics Association (IEA), 2000). The objective of HFE is, therefore, to produce multiple benefits for various stakeholders, such as workers and customers (Dul et al., 2012). In the domain of health care, there should be benefits for both patients and health care workers. Therefore, I conceptualize health care safety as safety for patients and safety for health care workers. In this book, health care safety is broadly defined and encompasses multiple dimensions of safety, health, well-being, quality of (working) life and performance for patients (and their care partners) and health care workers.

The HFE discipline is about making things better for people. It is based on humanistic values and promotes empathy in its methods and applications. It can help to achieve the World Health Organization (WHO) goals for universal health coverage and access to high-quality and safe patient care by focusing on the design of care processes. Ensuring that work systems and care processes are well-designed will produce benefits for both patients and health care workers.

1.1 PATIENT SAFETY

Patient safety is often described in terms of adverse events and subsequent harm to patients. In this section, I describe various approaches to patient safety and specific domains of patient safety (i.e. medication safety, health care-associated infections (HAIs) and diagnostic safety). I also show how systems approaches (including SEIPS) have been used to analyze and improve patient safety. I discuss misunderstandings around the concept of error and provide a brief description of Safety-I and Safety-II. I conclude by emphasizing the need for a systems approach to patient safety, which is the core argument of this book.

DOI: 10.1201/9781003480310-1

1.1.1 Definitions of and Approaches to Patient Safety

Over the years, different definitions of and approaches to patient safety have been proposed. For a review of different terms used for patient safety (e.g. medical errors, iatrogenic disease, medical harm), see, for example, the first chapter of the book 'Patient Safety' by Vincent (2010).

In the influential report by the Institute of Medicine (IOM) published in 2000 'To Err is Human: Building a Safer Health System' (Kohn et al., 2000), 'safety is defined as freedom from accidental injury' (page 58). In the 2001 follow-up report 'Crossing the Quality Chasm: A New Health System for the 21st Century', the IOM describes six quality aims, including safety, which is defined as 'avoiding injuries to patients from the care that is intended to help them' (Institute of Medicine, 2001, p. 6). An updated definition is used by the US Agency for Healthcare Research and Quality (AHRQ) Patient Safety Network (also known as PSNet), where patient safety is defined as 'avoiding harm to patients from care that is intended to help them' (https://psnet.ahrq.gov/patient-safety-101) (AHRQ PSNet, 2024). The definition of Vincent (2010) also draws attention to 'adverse outcomes or injuries stemming from the process of healthcare' (page 31). These various definitions emphasize patient harm and associated adverse events as a key aspect of patient safety.

As described by Wears and Sutcliffe (2020) in their book 'Still Not Safe: Patient Safety and the Middle-Managing of American Medicine', the approaches and definitions of harm and safety in health care have changed and evolved over time. An example of this evolution is the definition and approach used by the WHO. In 2009, the WHO defined patient safety as 'the reduction of risk of unnecessary harm associated with healthcare to an acceptable minimum' (World Health Organization (WHO), 2009, p. 15). Interestingly, the definition evolved, and, in the 2021–2030 Global Patient Safety Action Plan, the WHO proposed a more comprehensive definition of patient safety (World Health Organization (WHO), 2021):

> A framework of organized activities that creates cultures, processes, procedures, behaviours, technologies and environments in health care that consistently and sustainably lower risks, reduce the occurrence of avoidable harm, make errors less likely and reduce the impact of harm when it does occur.
>
> (page vii)

This long definition calls attention to patient safety as an outcome, as well as to the various factors and activities that positively and negatively influence patient safety and patient harm. This is in line with the SEIPS approach described in this book, which focuses on improving the design of work systems and care processes in order to reduce harm and produce benefits for both patients and health care workers.

1.1.2 Domains of Patient Safety

Patient safety is often identified with particular domains, such as medication safety, health care-associated infections (HAIs), diagnostic safety, surgical complications (e.g. postoperative pulmonary embolism (PE) and deep vein thrombosis), wrong-site surgeries, pressure ulcers, patient falls, missed care, and failure to rescue (see Table 1.1 for examples of patient safety issues). In the following sections, I discuss three specific patient safety domains: medication safety, HAIs and diagnostic safety. I selected these domains as they represent frequent and harmful patient safety issues, and, in some of my work, I have conducted research to address these three patient safety domains. I mention several of my research projects as well as research performed by other investigators. I briefly describe each domain, and then focus on specific aspects of the patient safety domains that relate to HFE and the Systems Engineering Initiative for Patient Safety (SEIPS) approach.

I discuss patient safety domains separately, but it is important to realize that these domains are inter-related. As argued by Pronovost and Bo-Linn (2012), 'The "siloing" of preventing patient harm is inefficient' (page 769); often, patients experience more than one type of harm. Therefore, a systems approach is needed to understand patient safety from a holistic systems perspective; this is a key argument in this book.

1.1.3 Medication Safety

Medication safety remains a critical domain of patient safety for hospitalized patients (Bates et al., 2023) as well as in primary care and long-term care (Agency for Healthcare Research and Quality (AHRQ), 2024). The American Society of Health-System Pharmacists defines medication safety 'as the freedom from accidental injury due to medical care or medical errors during the medication-use process' (Larson & Saine, 2013, p. 5). A frequently used approach to medication safety is based on the

TABLE 1.1
Sample Domains of Patient Safety

Domains of Patient Safety	Examples
Medication safety	Preventable adverse drug events (ADEs) Medication errors at stages of the medication use process: ordering/prescribing, dispensing, administration and monitoring
Health care-associated infections (HAIs)	Ventilator-associated pneumonia (VAP) Central line-associated bloodstream infections (BSIs) *C. difficile* infections (CDI) Surgical site infections
Diagnostic safety	Wrong diagnosis Delayed diagnosis
Surgery-related issues	Surgical complications Wrong site surgeries
Health care safety issues	Pressure ulcers Inpatient falls Missed care Failure to rescue Venous thromboembolism (VTE), including pulmonary embolism (PE)
Maternal health and safety	Safety of care for premature infants Complications during and following pregnancy and childbirth

research by Bates, Leape and colleagues. They define an adverse drug event (ADE) as 'an injury resulting from medical intervention related to a drug' (Bates et al., 1995, p. 29). Preventable ADEs are a subset of ADEs that are 'due to an error or preventable by any means currently available' (Bates et al., 1995, p. 30). In a landmark study of 4,031 admissions to two hospitals over a 6-month period, Bates et al. (1995) found 6.5 ADEs and 1.8 preventable ADEs per 100 hospital admissions. An example of a preventable ADE is the case of a patient with diabetes that has an order of NPO (nothing by mouth), but their insulin dose was not reduced, and the patient suffered significant reactions. This approach to assessing medication safety has led to numerous studies in various populations, e.g. pediatrics (Kaushal et al., 2001); in various care settings, e.g. intensive care units (ICUs) (Cullen et al., 1997); and around the world, e.g. in Japan (Morimoto et al., 2011).

Research on medication safety has often focused on identifying medication errors and challenges at different stages of the medication use process: ordering/prescribing, dispensing, administration and monitoring. Whereas early research quantified medication errors at each stage (Bates et al., 1995), it is important to recognize that these stages are not independent. An error at one stage can lead to an error at another stage; this is called propagation of errors (Carayon, Ju, et al., 2015). For instance, an error at the preparation stage (e.g. 2 pills in a package instead of 1 pill) could lead to a medication overdose if both pills are administered. In line with research on error detection and recovery (Nyssen & Blavier, 2006), an error at one stage can be identified and recovered at a subsequent stage (Carayon et al., 2016). For instance, a nurse recognized that there were 2 pills in the package, instead of 1 pill as prescribed; after verification, the nurse administered the prescribed single pill. Kuitunen et al. (2024) show linkages between errors in the medication management process for hospitalized children, such as errors in prescribing that lead to medication administration errors. Therefore, it is important to look at medication safety in a longitudinal, temporal manner, including activities that lead to the detection and recovery of medication errors (Carayon et al., 2016; Nyssen & Blavier, 2006).

In addition to linkages between various stages of the medication use process, medication safety can be enhanced or jeopardized in care transitions (World Health Organization (WHO), 2019b). Therefore, it is important to understand the medication management process when a patient is transferred, for example, between different care settings such as transfer from the hospital to a long-term care facility (Werner et al., 2017). Understanding how medication safety is enhanced or jeopardized over time corresponds with the idea of patient journey. As the patient transitions between different care settings, their journey is accompanied by multiple individuals and organizations; this is the core of the SEIPS 3.0 model of patient journey (Carayon, Wooldridge, et al., 2020). Because the patient is constant in the journey, efforts at improving medication safety during the journey and across care transitions should genuinely involve patients and their representatives and care partners (Grimes, 2021); this is further addressed when discussing the SEIPS 3.0 model of patient journey (Carayon, Wooldridge, et al., 2020) in Chapter 3 on SEIPS.

Interestingly, one of the earliest studies on medication safety was performed by the well-known HFE expert, Al Chapanis, and a colleague (Chapanis & Safrin, 1960; Safren & Chapanis, 1960a, 1960b). Using the critical incident technique (Flanagan, 1954) in a study performed in a 1,100-bed hospital over 7 months, Chapanis and his colleague identified 178 medication errors, including, for example, the wrong patient receiving or almost receiving a medication or the wrong medication dose. They also identified various contributing factors to medication safety, such as verbal or written miscommunication. Multiple studies have shown that medication safety is influenced by various system factors (Flynn, 2012). Studies have used the SEIPS model to gather data on factors contributing to medication safety; for instance, in an Ethiopian hospital's emergency department (ED) (Abebe & Chui, 2018), in community pharmacies in Saudi Arabia (Al Juffali et al., 2019), and in an outpatient mental health institution (Steele et al., 2018). An analysis of anticoagulation medication safety events shows that all work system elements of the SEIPS model are involved in incidents (Leon et al., 2024). Interestingly, the analysis shows that incident reports provided information on both work system barriers and facilitators, i.e. factors that lead to adverse events as well as factors that enhance medication safety. For instance, having contact details for the specialist anticoagulation team (*organization element of the work system*) supported communication within the health care team and could help to prevent the adverse event of inappropriate anticoagulation prescription.

1.1.4 Health Care-Associated Infections (HAIs)

Another major domain of patient safety is HAIs, such as ventilator-associated pneumonia (VAP), central line–associated bloodstream infections (BSIs), *C. difficile* infections (CDIs), and surgical site infections (SSIs). HAIs are frequent, harmful and costly (Eyal et al., 2013; Forrester et al., 2022; Guest et al., 2020). In a study led by the US Centers for Disease Control and Prevention (CDC), using 2011 data from 183 hospitals, Magill et al. (2014) found that 4% of hospitalized patients experience an HAI. Four years later, in 2015, fewer patients had HAIs (3.2%). Whereas improvements in HAI prevalence and incidence have been observed, in particular in the United States (Kwon et al., 2024; Magill et al., 2014), progress is still needed, especially for VAP, CDIs and SSIs (Barker, Ngam, et al., 2017; Horgan et al., 2024; Papazian et al., 2020; Seidelman et al., 2023).

Various infection control and prevention interventions have been developed and implemented to prevent, reduce and eliminate different types of HAIs. Many interventions take the form of 'bundles', i.e. a set of subprocesses and practices that are put together to improve patient safety. For example, Keenan et al. (2014) described a bundle aimed at preventing SSIs for colorectal surgery patients. The bundle includes a set of practices for each stage of the surgical process:

1. preoperative subprocess: e.g. chlorhexidine shower and mechanical bowel preparation with oral antibiotics,
2. operative subprocess: e.g. a dedicated wound closure tray and limited operating room traffic and
3. postoperative subprocess: e.g. daily washings of incisions with chlorhexidine.

The bundle also includes activities that are common across stages, such as patient education. Another example of a bundle is the VAP bundle developed by a European group, which includes activities for the diagnosis of VAP (i.e. early chest X-ray with specialist interpretation within 1 hour, immediate reporting of respiratory secretions Gram staining) and activities for the treatment of VAP (i.e. immediate treatment, empiric therapy based on local pathogens and risk factors, de-escalation, assessment of response within 72 hours and short therapy duration if feasible) (Rello et al., 2011).

The well-known bundle of Pronovost et al. (2006) in the Keystone Michigan Project aimed at eliminating catheter-related BSIs. A total of 103 ICUs in the US state of Michigan participated in the study and reported data on catheter-related BSIs. The bundle consisted of five evidence-based procedures:

1. hand washing,
2. using full-barrier precautions during the insertion of central venous catheters,
3. cleaning the skin with chlorhexidine,
4. avoiding the femoral site if possible and
5. removing unnecessary catheters.

The implementation of the bundle was supported by training of physicians, a central-line cart with necessary supplies, a checklist of infection prevention practices with nurse empowerment to stop the procedure if the practices were not followed and questioning the need for catheters during bedside rounds (Berenholtz et al., 2004; Pronovost et al., 2006). The intervention led to major decreases in catheter-related BSIs per 1,000 catheter-days: median dropped from 2.7 to 0 after 3 months and sustained at 0 during the 18 months of follow-up (Pronovost et al., 2006). The reduced rates of infections were sustained for another 18 months after the initial 18 months of post-implementation follow-up of the intervention (Pronovost et al., 2010). In addition, hospital mortality for patients in the ICUs that implemented the bundle was lower as compared to a control group of nonparticipating ICUs (Lipitz-Snyderman et al., 2011).

A retrospective analysis of the Keystone Michigan Project showed that, in order to reduce catheter-related BSIs, it was necessary to go beyond just the bundle; the bundle is embedded in a complex systemic program with technical, equipment, organizational and cultural characteristics (Dixon-Woods et al., 2011). The bundle can actually be described using the work system model (Carayon, 2009; Smith & Carayon-Sainfort, 1989). Each element of the bundle is supported by an entire work system and its elements. For instance, 'cleaning the skin with chlorhexidine' is a *task* that requires several *tools* (e.g. chlorhexidine product, cloth) and takes place in a constrained *physical environment* (e.g. patient room with limited space, noise from devices and interruptions). The nurse performing the *task* needs training on how to effectively use chlorhexidine to clean the skin; the training is provided by the *organization*. This short example shows how an intervention such as the bundle of the Keystone Michigan Project can be conceptualized as a work system intervention.

HAI bundles are composed of various activities performed by multiple individuals, and these activities are embedded in a work system. Let us look at a *C. difficile* bundle as an example. *Clostridioides difficile (C. difficile)* infection (CDI) is a critical patient safety issue and can be largely prevented by infection control practices, which can be bundled together. CDI bundles can have a significant impact, but their implementation remains challenging (Barker, Ngam, et al., 2017). To address these challenges, I propose to conceptualize bundles as a systemic change that affects the work done by various individuals and may produce barriers or facilitators at the sharp end of patient care; these work system barriers and facilitators, in turn, can affect the effectiveness of the bundle.

In a series of studies (Musuuza et al., 2019; Ngam et al., 2017; Yanke et al., 2018), our interdisciplinary team examined the implementation and use of a CDI bundle in multiple settings from the perspectives of various stakeholders directly involved in the bundle implementation. The CDI bundle consisted of the following five components or subprocesses (Musuuza et al., 2019):

1. diagnostic testing,
2. empiric isolation for patients with suspected CDI,
3. contact isolation,
4. hand hygiene and
5. environmental disinfection.

Using a total of eight focus groups with physicians, nurses, health technicians and environmental services staff in two hospitals (Musuuza et al., 2019; Ngam et al., 2017; Yanke et al., 2018), we identified multiple barriers and facilitators for each of the subprocesses. For example, for nurses, a major barrier was associated with the *task element of the work system*: workload or the additional effort required to don and doff gloves and gowns (subprocesses of empiric isolation and contact isolation) (Ngam et al., 2017). The study by Musuuza et al. (2019) compared the experience and use of the CDI bundle by physicians, nurses and environmental staff services and found common, unique and conflicting barriers and facilitators across the three groups. For instance, all three groups experience the following work system barriers: poor layout of the *physical environment* (e.g. inconvenient location of sinks) and inconsistencies in CDI documentation (*interaction between the task and technology elements of the work system*). Unique facilitators included algorithmic approaches used by physicians that helped with the timely diagnosis of CDI (*technology element* that supports the performance of the diagnostic *task*). Conflicting barriers between the three groups were related to different objectives. For example, physicians and nurses needed quick access to rooms for placing a patient while the environmental staff needed sufficient time to disinfect the room (*interaction between the task, physical environment and organization elements of the work system*). This study highlights the importance of considering multiple stakeholder groups and perspectives in designing patient safety improvements. This is further discussed in Chapter 4 on Human-Centered Design (HCD).

Antibiotic resistance is a major problem given the complexity of preventing and treating HAIs. Therefore, antibiotic stewardship or 'the effort to measure and improve how antibiotics are prescribed by clinicians and used by patients' (https://www.cdc.gov/antibiotic-use/hcp/core-elements/index.html#toc) has emerged as another HAI-related intervention. Antibiotic stewardship programs have been largely implemented in hospitals (CDC, 2019). Like any intervention in health care, antibiotic stewardship represents a change in how work is done and care is given, and, therefore, should be conceptualized as a work system intervention. As an example, let us look at an antibiotic stewardship intervention aimed at reducing the prescription of antibiotics in the fluoroquinolone (FQ) class for critical care patients. FQ is a broad-spectrum antibiotic, which unfortunately increases the risk of *C. difficile*. In a study aimed at reducing FQ prescription for critical care patients in two hospitals, we implemented an antibiotic stewardship program with pre-prescription authorization and computerized decision support using a best practice alert in the electronic health record (EHR) (Carayon, Thuemling, et al., 2021). In both hospitals, the intervention led to decreased use of FQ. The SEIPS model was used to describe the intervention in each of the two hospitals, thereby helping to understand the intervention as a work system change, which involved all elements of the work system. This application of SEIPS to describe an intervention is further discussed in Chapter 3 on SEIPS.

BOX – INTERVENTIONS AIMED AT HAIS = WORK SYSTEM INTERVENTIONS

Interventions to prevent, reduce or eliminate HAIs often take the form of bundles or involve programs such as antibiotic stewardship. These interventions involve changes in how work is done and care is given; therefore, they are work system interventions as multiple work system elements and their interactions change. The work system interventions change:

- Who does the work? [*people*]
- What people do? [*tasks*]
- What *tools and technology* are used to do the work?
- Where the work is done? [*physical environment*]
- How the work is done? [*organization*]

Actually, all patient safety improvement activities can be conceptualized as work system interventions.

There are many ways that the HFE discipline contributes to analyzing, reducing, eliminating and preventing HAI. For instance, researchers at the Armstrong Institute for Patient Safety at Johns Hopkins University in the United States have described the role of HFE in preventing and reducing HAIs by identifying barriers and facilitators in the use of personal protective equipment (PPE) (Gurses et al., 2019) and in the room cleaning process for operating rooms (Xie et al., 2024) and patient rooms (Xie et al., 2018). These studies highlight the importance of conceptualizing HAI-related interventions as work system changes. Such SEIPS-based analysis can help to improve the design, implementation and effectiveness of HAI-related interventions.

1.1.5 Diagnostic Safety

The 2015 publication of the Institute of Medicine (IOM) report 'Improving Diagnosis in Health Care' (Balogh et al., 2015) brought long-awaited attention to diagnostic safety. According to the report, 'most people will experience at least one diagnostic error in their lifetime, sometimes with devastating consequences' (page S–1). The report proposes a patient-centered definition of diagnostic error: 'the failure to (a) establish an accurate and timely explanation of the patient's health problem(s) or (b) communicate that explanation to the patient' (page 1–5).

A major contribution of the report is its focus on the diagnostic process. Diagnosis is not one event or one decision made by a physician; it unfolds over time and involves multiple individuals collaborating and working together, including the patient. In this IOM report (Balogh et al., 2015), the diagnostic process is conceptualized as a series of activities performed by various members of the diagnostic team, including the patient and, potentially, their care partner(s). The process begins when a patient experiences a health problem and attempts to engage with a health care system. The core of the diagnostic process includes the following interacting, iterative cognitive activities: information gathering, information integration and interpretation and working diagnosis. These activities take place in a work system (Smith & Carayon-Sainfort, 1989), which can either enhance or hinder the performance of the cognitive activities. The process continues with the communication of the diagnosis to the patient and follow-up treatment, if necessary. Outcomes of the overall diagnostic process could be positive (i.e. accurate, timely diagnoses) or negative (i.e. diagnostic errors or near misses) and then affect patients and the overall health care system. In a 2024 report on 'The Economics of Diagnostic Safety', the Organisation for Economic Co-operation and Development (OECD) (2024) adopted the IOM model to describe the diagnostic process, not as a one-off activity, but as an iterative and complex process where multiple people interact.

According to the IOM conceptual model of diagnostic safety (Balogh et al., 2015), the diagnostic process is embedded in the work system; therefore, one can look at factors within the work system as targets for potential diagnostic safety improvement interventions. For instance, using the SEIPS model (Carayon, Hundt, Karsh, et al., 2006), Sarkar et al. (2014) identified numerous factors contributing to delayed or missed diagnoses in outpatient care. Based on data from six focus groups with 25 physicians, they identified multiple factors within each of the work system elements, such as inefficient referral and organizational culture in the *organization element.* Others have also advocated for a systems approach, such as SEIPS, to address diagnostic safety (Ladell et al., 2025).

Early research on diagnostic safety differentiated between individual (cognitive) factors and system-related factors that contribute to diagnostic errors (Graber, 2013; Graber et al., 2005). A study by Graber et al. (2005) analyzed 100 diagnostic errors and identified two separate lists of contributing factors: (1) system-related factors (e.g. faulty or miscalibrated testing equipment, communication failures, lack of oversight of medical trainees) and (2) cognitive factors (e.g. insufficient knowledge, failure to collect required information from patient, missing symptoms). This distinction between the individual and the system does not fit with models and frameworks developed by the HFE community; the individual is embedded in and is part of the work system. Cognitive activities are not stand-alone activities; they are influenced by the work system in which the individual works. For instance, a physician may miss a symptom (cognitive factor) because of interruptions (characteristic of the work system). A study by Gupta et al. (2018) confirms that various system factors affect cognitive activities involved in the diagnostic process. For example, as described above in the conceptual model on diagnostic safety proposed by the IOM (Balogh et al., 2015), a key cognitive activity is gathering and sharing of patient-related information in a timely manner by nurses and physicians. When nurses are not systematically involved and integrated in bedside rounding for hospitalized patients, key clinical information may be missed, which could lead to a wrong diagnosis or a delayed diagnosis. This example shows how the *organization element of the work system* (participation of nurses in bedside rounding) affects the cognitive activity of information gathering.

The conceptual model proposed in the 2015 IOM report (Balogh et al., 2015) clearly shows how the cognitive activities of information gathering, information integration and interpretation and working diagnosis are embedded in a work system. The IOM model incorporates HFE knowledge on information processing stages (Parasuraman et al., 2000) and the HFE-based work system model (Carayon, 2009; Smith & Carayon-Sainfort, 1989). According to Hutchins' (1995, 2001) concept of distributed cognition, information is shared and propagated through a system (e.g. other people, the physical environment, artifacts); therefore, we need to look at cognitive activities as they are related to the work system in which they occur.

1.1.6 Patient Safety, Human Error and Error

The discussion on patient safety has sometimes focused on reducing errors as a goal. For example, the updated definition of patient safety by the WHO (2021) includes 'making errors less likely' as an objective of patient safety. The 2023 report by the US President's Council of Advisors on Science and Technology (PCAST) on patient safety discusses preventable and avoidable (medical) errors (President's Council of Advisors on Science and Technology, 2023). The focus on reducing, preventing or eliminating errors is aligned with what Dekker (2014) called the 'old view' on human error when the error is the problem and, therefore, a target for intervention. We need to go beyond the focus on error as the main contributor to safety and understand the work system and its characteristics that influence errors, create obstacles or foster safety and performance; this systems approach helps to identify a wider range of possible interventions for improving patient safety (Catchpole & Wiegmann, 2012).

In discussions on patient safety, the term 'error' is often not used in light of the state of science on 'human error' (Read et al., 2021; Reason, 1997; Wears & Sutcliffe, 2020). In some publications, the term 'error' is not used to refer to 'human error', as described and discussed in the HFE literature and by experts such as James Reason (1997). This adds to the confusion and may limit our progress in improving patient safety as models and theories of human error are not applied. The confusion about terminology also applies to HFE. For instance, the term '*human factors* has been described as similar to 'error promoting conditions' (Graber et al., 2024).

Human Factors and Ergonomics (HFE) is a scientific discipline aimed at improving work system design for both performance and well-being (International Ergonomics Association (IEA), 2000); it is not about eliminating human errors (Russ et al., 2013). When using various terms and concepts, in particular error, we need to carefully define the terms and concepts and their conceptual basis (Le Coze, 2022).

When patient safety focuses on errors, it is often about things that are done, but go wrong, e.g. wrong medication given or wrong diagnosis. It is important to also think about patient harm when something is not done. For instance, McGlynn et al. (2003) showed that about half of the participants (55%) in a large-scale study across the United States did not receive the care recommended; this was true for preventive care (55%; e.g. mammographic screening for breast cancer not performed), acute care (54%; e.g. prophylactic antithrombotic drugs not given to hip surgery patients) and care for chronic conditions (56%; e.g. not advising smokers with chronic obstructive pulmonary disease to quit smoking). A similar study published in 2016 showed comparable results on the quality of outpatient care (Levine et al., 2016): recommended medical treatment and recommended cancer screening remained suboptimal. Not receiving care can be considered a safety issue as it can lead to patient harm. For instance, if a patient does not get the required preventive screening, harmful conditions may not be diagnosed in a timely manner. Therefore, we need to look at whether the patient is actually able to access and interact with health care organizations. This first step is part of the conceptual model proposed in the IOM report on diagnostic safety (Balogh et al., 2015): when the patient experiences a health problem, is s/he able to engage with the health care system? If barriers exist in the patient–health care system interaction, then the diagnostic process cannot unfold, and the patient may be ultimately harmed as a disease may not be diagnosed and treated in a timely manner.

Some international reports on patient safety do not use the term error, such as the 2023 UK Imperial College report on global patient safety (Illingworth et al., 2023). According to some authors, the focus on human error in safety (including patient safety) is not welcome (Braithwaite et al., 2015; Dekker, 2011; Wears & Sutcliffe, 2020). Wears and Sutcliffe (2020) argue that the discussion on patient safety was wrongly reframed as medical errors. The patient safety approach focused on human error tends to draw attention to the individual behavior of health care workers and errors committed during various actions. This tension between focusing on the individual or focusing on the rest of the system and care processes was debated at a 2014 conference organized by the National Patient Safety Foundation (McTiernan et al., 2015). Are we focusing too much on the individual who made the error or an individual who did not follow a protocol? Are we looking at the specific situation where the error or problem occurred and the work system that influenced the situation? We need to go beyond the focus on error and understand the hazards and obstacles as well as the facilitators in the work system (Ladell et al., 2025).

The HFE discipline and safety sciences have evolved significantly and point to the systems approach as the effective means to safety improvement (Dekker, 2011; Read et al., 2021). In addition, I should re-emphasize that the individual is part of the system. We should focus on the complex systems in which people work and care processes are embedded with the aim of improving the design of work systems and care processes for patient safety. This is what I do in this book as I describe the SEIPS-based systems approach to health care safety.

1.1.7 Safety-I and Safety-II

Safety thinking and approaches have been described and categorized in many different ways, including Safety-I and Safety-II. According to Hollnagel (2014), Safety-I focuses on things that go wrong, aims at reducing adverse outcomes (e.g. accidents, incidents, near misses) and is a reactive approach to safety management. Hollnagel contrasts Safety-I with Safety-II, which focuses on things that go right and the strategies people use when facing problems and disturbances; Safety-II is a proactive approach to safety management. Safety-II has received significant interest and attention in health care (Iflaifel et al., 2020; Mannion & Braithwaite, 2017; Woodward, 2019). The descriptions of Safety I and Safety II and their different underlying principles and assumptions have been criticized (Cooper, 2020; Lawton, 2018; Leveson, 2020; Sujan, 2024), and challenges to the application of Safety-II to health care remain to be explored (Verhagen et al., 2022). Does Safety-I exist or is it a strawman concept (Leveson, 2020; Sujan, 2024)? What does Safety-II offer to improve patient safety (Sujan, 2024)?

Some researchers have proposed to combine Safety-I and Safety-II; for instance, in an analysis of patient safety in virtual encounters between patients and primary care (Payne et al., 2024) and in an analysis of procedural violations in community pharmacies (Jones et al., 2018). Therefore, it may not be necessary to oppose Safety-II to Safety-I and other approaches to safety. On the contrary, we need a requisite variety of perspectives to address the complex safety problems in health care (Lawton, 2018). I agree with many experts and commentators that we need various approaches that provide different insights and guidance to deal with complex health care safety problems (Lawton, 2018; Verhagen et al., 2022; Vincent & Amalberti, 2016). In this book, I focus on SEIPS, but also describe multiple examples of how to combine SEIPS with other approaches, models and methods.

Safety-II calls attention to the complexity of health care systems (Braithwaite et al., 2015; Hollnagel et al., 2015). This is important as it fits with systems thinking in patient safety and the SEIPS-based systems approach described in this book. Safety-II aims to understand the actual work (i.e. 'work-as-done' [WAD]) that occurs in complex health care systems; this is critical. WAD contrasts with procedures or 'work-as-imagined' (WAI). Safety-II has led to the development of the Functional Resonance Analysis Method (FRAM) to study WAD in practice (Hollnagel, 2012b). The FRAM has been applied to analyze the double checking of injectable medication administration (Schutijser et al., 2019) and to model transitional care of older adults during and after a hospitalization (O'Hara et al., 2020). FRAM has been promoted as a Safety-II method to analyze WAD. In Chapter 4 on Human-Centered Design and SEIPS in Action, I discuss the concepts of WAD and WAI that are described in the Safety-II and Resilience Engineering literature and highlight the rich HFE toolbox of methods for analyzing work systems.

Combining multiple methods and approaches will provide critical insights for understanding and improving health care safety. For example, Pickup et al. (2017)

combine FRAM and SEIPS to analyze safety in the blood collection process. A challenge with Safety-II and FRAM is the design of safety improvements and work system interventions. We can learn a lot from FRAM analyses; a challenge is to translate the insights from the analyses into interventions (Lawton, 2018; Sujan, 2024; Sujan, Lounsbury, et al., 2024; Verhagen et al., 2022).

An important contribution of Safety-II and Resilience Engineering is the concept of strategies that individuals develop and use when facing disturbances. Resilient strategies include monitoring, responding, anticipating and learning (Hollnagel, 2011; Hollnagel & Woods, 2006). Those strategies have been found among primary care physicians who coordinate and prioritize referrals to specialists (Arcuri et al., 2020) and in a study of individual, team and organizational responses to patient safety incidents (Leon et al., 2024). An extensive taxonomy of adaptive strategies was developed based on a literature review (Page et al., 2024); the four resilient strategies represent one type of adaptation to disturbances and pressures. We need to understand the conditions (i.e. work system) that foster or hinder these individual behaviors or strategies (Leveson, 2020); this could be done with SEIPS. The systems approach to patient safety, therefore, needs to ensure that work systems and care processes are designed to foster individual strategies for dealing with hazards, pressures, unexpected events and other disturbances.

1.1.8 Systems Approach to Patient Safety

A systems approach to patient safety means looking at the various work system factors that contribute to patient safety and other outcomes; this involves a deep analysis of the actual work done by health care staff, patients and care partners. In Sections 1.1.3 to 1.1.7, I mentioned systems approaches for specific patient safety domains. Various systems approaches and systems thinking are discussed in depth in Chapter 2 on Systems Thinking in Health Care Safety.

For several years, the US Agency for Healthcare Research and Quality (AHRQ) issued reports on patient safety practices (Shekelle, Wachter, et al., 2013; Shojania et al., 2001). These reports produced lists of patient safety practices, and the lists were getting longer over time, therefore, requiring the need to prioritize and limit the number of key patient safety practices to recommend to health care organizations (Shekelle, Pronovost, et al., 2013). For instance, the second report on 'Making Healthcare Safer' strongly encouraged 10 patient safety practices (Shekelle, Pronovost, et al., 2013):

1. checklists to prevent operative and postoperative events,
2. bundles that include checklists to prevent central line–associated BSIs,
3. interventions to reduce urinary catheter use, including catheter reminders, stop orders, or nurse-initiated removal protocols,
4. bundles that include head-of-bed elevation, sedation vacations, oral care with chlorhexidine and subglottic suctioning endotracheal tubes to prevent ventilator-associated pneumonia (VAP),
5. hand hygiene,
6. the do-not-use list for hazardous abbreviations,
7. multicomponent interventions to reduce pressure ulcers,

8. barrier precautions to prevent HAIs,
9. use of real-time ultrasonography for central line placement and
10. interventions to improve prophylaxis for venous thromboembolism (VTE).

Another 12 patient safety practices were encouraged, such as multicomponent interventions to reduce falls, team training and computerized provider order entry (CPOE). Many of the patient safety practices are specific to a domain of patient safety. For instance, practices #2, #3 and #4 above target specific HAIs: central line BSIs for practice #2, catheter-related HAIs for practice #3 and VAP for practice #4. Other patient safety practices are not specific to a patient safety domain, such as practice #1 about checklists and practice #6 about not using hazardous abbreviations.

This approach of focusing on specific patient safety practices for specific patient safety domains does not recognize that often patients experience multiple types of harm (Pronovost & Bo-Linn, 2012). In addition, some of the same work system factors contribute to more than one domain of patient safety, and any patient safety practice is embedded in a broader work system. An important example of work system factor is the workload experienced by nurses. High workload of nurses has been associated with problems concerning medication safety among ICU patients (Carayon & Gurses, 2005; Tarnow-Mordi et al., 2000), HAIs (Stone et al., 2004) and inpatient falls for hospitalized patients (Krauss et al., 2005). Therefore, high workload is a common factor contributing to multiple domains of patient safety. A systems approach emphasizes the need to look at these multiple connected safety outcomes influenced by high workload and understand the work system factors that in turn influence workload. High workload is often assessed as the nurse–patient ratio, but this is not sufficient. We need to look at the work system factors that contribute to nursing workload, which can then lead to multiple safety problems (Carayon & Gurses, 2005). Therefore, in addition to looking at multiple safety outcomes, a systems approach would take a step back and identify the various factors that contribute to nursing workload, such as inadequately stocked patient rooms (related to the *organization element of the work system*) or poorly designed *physical environment* (Gurses et al., 2009). This systems approach to workload would also need to consider the well-known negative impact of workload on occupational health and safety, job stress and burnout of health care workers. This example of workload is an important example that can be expanded to other work system characteristics. The key idea is that multiple patient safety domains are influenced by some of the same system factors; therefore, health care organizations should not only address specific patient safety domains but also look at the systemic factors that influence multiple patient safety domains. We also need to look at safety for health care workers.

1.2 HEALTH CARE WORKER SAFETY

Health care worker safety is the term that I broadly use to encompass occupational health and safety, worker well-being and quality of working life. Safety of health care workers has physical and psychological dimensions and can be described in positive and negative terms; see Table 1.2 for examples. From a physical perspective, safety of health care workers covers issues of work-related musculoskeletal disorders

TABLE 1.2
Multiple Dimensions of Health Care Worker Safety – Examples

Physical Dimensions of Health Care Worker Safety	Psychological Dimensions of Health Care Worker Safety
Occupational health and safety	Job stress
Occupational infections	Burnout
Work-related musculoskeletal disorders (WRMDs)	Fatigue
Sharps and needlestick injuries	Mental ill health
	Quality of working life

(WRMDs), sharps injuries, falls, infections, etc. From a psychological perspective, health care workers report high stress, fatigue and burnout, and poor quality of working life (e.g. job dissatisfaction), and experience mental ill health problems, such as anxiety and depression.

1.2.1 Occupational Health and Safety: Injuries and Illnesses

Health care workers are exposed to multiple health and safety hazards, including biological hazards, unsafe patient handling, workplace violence and work pressure, which lead to worker injury and harm, such as infections, physical injuries and ill-health (De Bienassis et al., 2021; Loeppke et al., 2017). For instance, health care workers are at high risk of sharps injuries and the related consequence of infections (Elseviers et al., 2014). Across nine studies performed around the world, on average per year, 3.7% health care workers, in particular nurses, experienced needlestick injuries, with potential transmission of infections from patients to workers (Elseviers et al., 2014); this demonstrates a link between health care worker safety and patient safety.

Data from the US Bureau of Labor Statistics show that the most common nonfatal injuries and illnesses of hospital workers are overexertion and bodily reaction (e.g. injuries from moving or lifting patients) and falls–slips–trips (Dressner, 2017). The nursing staff is at high risk of these injuries. For instance, the rate of WRMDs is higher for nurses than for any other occupation (Dressner & Kissinger, 2018). Several reviews confirm the high risk of WRMDs for nurses and nursing aides, in particular in the back, shoulders and neck (Davis & Kotowski, 2015; Ellapen & Narsigan, 2015). Across a total of 132 studies of nurses in hospitals, long-term care and home health care, the average lifetime prevalences were 65% for low-back pain and 54% for shoulder pain; the average annual prevalences were 55% for low-back pain, 44% for shoulder pain and 42% for neck pain (Davis & Kotowski, 2015).

Both physical and psychosocial work factors influence WRMDs (Carayon et al., 1999), in particular among health care workers (Bernal et al., 2015; Reme et al., 2014). For instance, a prospective study of 2,080 health care workers (e.g. nurses, occupational and physical therapists, doctors) from 17 Danish hospitals identified the following factors contributing to back injuries: a large number of patient handling activities (physical work factor) and poor collaboration between colleagues (psychosocial work factor) (Andersen et al., 2019). In addition, when workers experienced a back injury, they often reported lacking assistive devices such as sliding sheets or ceiling lifts. The lack of assistive devices has both physical and psychosocial aspects. If a nurse does not use an assistive device, she may overexert herself and experience a back injury (physical aspect). The nurse may not have easy access to the device; for example, the device may not be conveniently located (physical aspect) or the nurse may not have the appropriate training to safely use the device (psychosocial aspect). Therefore, both physical and psychosocial characteristics of work are important to consider when addressing occupational injuries, such as WRMDs. We need a systems approach to address the multiple physical and psychosocial work factors that contribute to health and safety hazards experienced by health care workers.

1.2.2 Psychological Aspects of Health Care Worker Safety

Health care work has also been associated with high stress, fatigue, burnout and mental ill-health and low quality of working life (Bernal et al., 2015; De Bienassis et al., 2021; Loeppke et al., 2017; NASEM, 2019). According to the US National Academies' report on clinician burnout, between 35% and 54% of nurses and physicians have substantial symptoms of burnout (NASEM, 2019). In an online survey of 745 hospital nurses, 50% of them experienced chronic fatigue (Barker & Nussbaum, 2011). A meta-analysis indicates that resident physicians (i.e. physicians-in-training) are at high risk of depression (summary estimate of the prevalence of depression: 29%; ranging from 21% to 43%) (Mata et al., 2015). These multiple studies point to the need to address various psychological aspects of health care worker safety.

A major report by the US National Academies was published in 2019: 'Taking Action Against Clinician Burnout: A Systems Approach to Professional Well-Being' (NASEM, 2019). Burnout is a work-related problem, as supported by WHO that includes burnout in the International Classification of Diseases (ICD-11) (World Health Organization (WHO), 2019a). Burnout is composed of three dimensions: exhaustion, depersonalization and inefficacy (Maslach et al., 2001). The 2019 National Academies' report provides evidence on clinician burnout (in particular, exhaustion) and its consequences for health care workers, health care organizations, patients and society. As indicated above, between 35 and 54% of physicians and nurses in the United States experience substantial symptoms of burnout, which have consequences for the clinicians themselves (e.g. risk of injuries, high depression) as well as for patients (e.g. poor communication between patients and clinicians, necessary care not performed, HAIs).

As indicated in the title of the 2019 National Academies' report ('Taking Action Against Clinician Burnout: A Systems Approach to Professional Well-Being'), we need to encourage a systems approach to reduce burnout and improve well-being. The report's conceptual framework defines the mechanisms by which work system factors (e.g. excessive workload, administrative burden, inadequate usability of technology, lack of job control) contribute to burnout and produce consequences for patients, clinicians, health care organizations and society. The conceptual framework also includes feedback loops of learning and improvement. Work system factors emerge from three levels of system:

1. the sharp end or frontline care delivery,
2. the health care organization and
3. the external environment.

The conceptual framework integrates the SEIPS work system model (Carayon, Hundt, Karsh, et al., 2006) and adopts several key features of systems thinking, in particular, interacting system levels and feedback loops. The National Academies' report emphasizes the need to go beyond individual-focused interventions and tackle the numerous work system factors that produce burnout. Unfortunately, this has not been sufficiently emphasized and many interventions to address clinician burnout continue to target wellness and individual resilience (Bock, 2024), whereas health care workers want their organizations to address work system issues, such as poor work environments and insufficient job control (Aiken et al., 2023).

The COVID-19 pandemic put the spotlight on the burnout and ill health of health care workers who often had to provide care with limited resources in high-stress environments. Physicians-in-training who were exposed to COVID-19 patients were more likely to report stress and burnout (Kannampallil et al., 2020). During the acceleration phase of COVID-19, emergency physicians were particularly at high risk and reported high levels of burnout and workplace anxiety (Rodriguez et al., 2020). Whereas the COVID-19 pandemic brought unique challenges (e.g. limited access to PPE), it heightened many of the stressors already present in health care work, such as high workload and issues with work–life balance (Carayon & Perry, 2021). For instance, a qualitative study of 32 Chinese health care professionals (e.g. physicians, nurses, pharmacists) identified a range of factors that influenced their well-being during the COVID pandemic (Zhang & Gu, 2025). The factors covered all the work system elements, such as challenges with usability of and access to PPE (*technology element of the work system*) and team cohesion (*organization element of the work system*). Multiple lessons learned have emerged in the post-pandemic literature (Knutsen Glette et al., 2023; Sinsky & Linzer, 2020; Wu et al., 2020), such as the need to reduce administrative burden and the associated workload by improving teamwork and eliminating unnecessary tasks. However, a lot of effort remains to improve working conditions for health care workers (Taylor et al., 2024).

1.2.3 Holistic Approach to Health Care Worker Safety

The COVID-19 pandemic has brought to the forefront not only problems of providing safe, high-quality care but also the issue of health care worker safety (Grant et al., 2023). During the pandemic, health care workers experienced fatigue, stress and burnout and were at high risk of HAIs. What happened during the COVID- 19 pandemic shows that health care workers often experience multiple negative outcomes.

The different dimensions and aspects of worker safety, as described in Table 1.2, are inter-related. For instance, a study by Reme et al. (2012) shows a link between psychological distress and musculoskeletal pain among a group of 1,572 patient care workers in two hospitals. Workers who reported high musculoskeletal pain within the past 3 months also reported high psychological distress (e.g. feeling sad, nervous, restless, hopeless).

From a holistic perspective, it is useful to refer to the concept of worker well-being as proposed by the US National Institute for Occupational Safety and Health (NIOSH) (Chari et al., 2018). The NIOSH worker well-being framework is composed of five domains (each with multiple sub-domains):

1. Workplace physical environment and safety climate: e.g. environmental conditions, physical surroundings, incivility, perception of safety climate.
2. Workplace policies and culture: e.g. salary and rewards, benefits, respect, fairness, organizational commitment to health, work-life integration.
3. Health status: physical health, mental health (e.g. stress, depression, anxiety), health-related behaviors, physical-mental-cognitive functionality, injuries.
4. Work evaluation and experience: job satisfaction, job security, meaning of work, autonomy, demands, emotions at work.
5. Home, community and society: life satisfaction, financial security, social support, community engagement.

The framework integrates work and nonwork factors that can affect various aspects and dimensions of worker health and quality of life. Improving worker well-being means that workers have the working conditions necessary to thrive and be safe. NIOSH produced a questionnaire to assess 16 dimensions across the five domains listed above (Chari et al., 2022).

1.2.4 Systems Approach to Health Care Worker Safety

In order to support, promote and improve health care worker safety, organizations should ensure that their workers are not exposed to physical and psychological hazards that can lead to accidents, injuries and diseases and affect well-being and health. Therefore, a systems approach to health care worker safety needs to consider multiple aspects of safety as well as multiple contributing factors to safety. A systematic review of 36 studies of nurses and nursing aids identified three categories of work-related factors and one category of individual factors that contribute to various safety outcomes, such as injuries, musculoskeletal discomfort and disorders, psychological distress and burnout (McCaughey et al., 2016):

1. Organizational structure and processes. For instance, working in nursing homes or critical care units increases the risk of occupational injuries. Shift work schedule and patient characteristics also influence worker safety.

2. Job characteristics. General workplace climate as well as the specific safety climate are key determinants of worker safety. Other important job characteristics include physical and psychological demands, and social support at work.
3. Safety programs and training. For instance, it is not only important to provide safety equipment to nurses and nursing aids but also to train them on how to use it and to consider the physical and organizational environment in which the equipment is used.
4. Individual characteristics of health care staff and their attitudinal outcomes (e.g. job satisfaction).

This systematic review confirms that a large variety of factors are at play in influencing health care worker safety. There is not one or a small set of factors that have a determinant impact on worker safety; this calls for a systems approach to worker safety, as recommended by experts and international organizations.

International organizations, such as the International Labour Organization (Alli, 2008) and WHO (2010), have advocated for healthy and safe workplaces that are adapted to workers' physical, physiological and psychological capabilities. WHO (2010) proposed a model of 'healthy workplaces' that defines four key areas for action:

1. The physical environment: e.g. noise, heavy lifting.
2. The psychosocial work environment: e.g. time pressure, lack of organizational support, 'toxic' organizational culture.
3. Personal health resources in the workplace: e.g. poor diet related to lack of access to healthy food.
4. Enterprise community involvement in order to improve the health of workers, their families and other community members: e.g. access to public transportation.

The WHO model of healthy workplaces is aligned with the SEIPS model described in this book as it addresses multiple dimensions of the work system: physical, psychosocial, organizational and community/external environment.

1.3 LINKAGE BETWEEN PATIENT SAFETY AND HEALTH CARE WORKER SAFETY

The various outcomes of patient safety and health care worker safety are related in multiple ways (see Figure 1.1):

1. The outcomes are positively related to each other.
2. The outcomes share similar contributing factors.
3. Actions or programs to improve one outcome may have a negative impact on the other outcome.

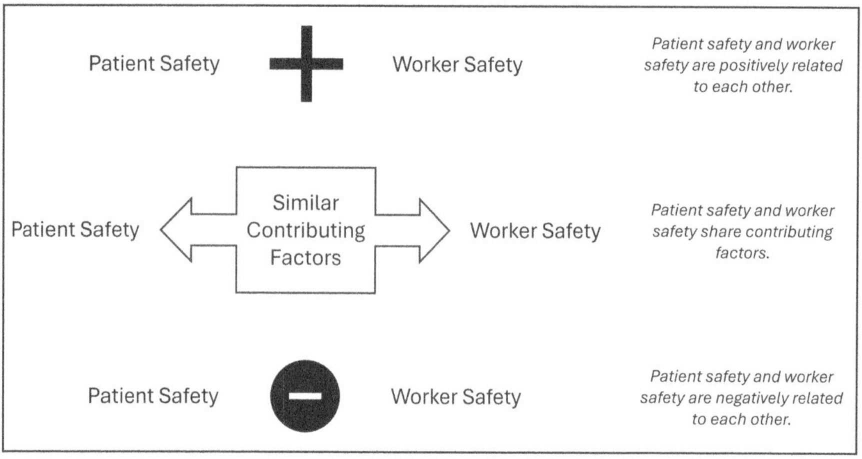

FIGURE 1.1 Linkages between patient safety and health care worker safety.

1.3.1 Patient Safety and Health Care Worker Safety are Positively Related to Each Other

Early on when patient safety emerged as a critical issue, a few experts called attention to the relationship between patient safety and health care worker safety (Lundstrom et al., 2002; The Joint Commission, 2012). However, for a long time, this synergistic relationship between patient safety and worker safety was largely ignored, even though the literature provides clear empirical evidence about the relationship between patient safety and worker safety.

A systematic review of health care staff well-being shows that, in the majority of the 46 studies included, both well-being and burnout are associated with patient safety (Hall et al., 2016). Poor well-being was measured as depression, anxiety, high stress and poor quality of life. Both subjective (e.g. self-perceived errors) and objective (e.g. event report frequency such as near misses) measures of patient safety were associated with low well-being and high burnout. These studies covered various groups of health care workers, mainly nurses and physicians. Another recent systematic review confirmed the relationship between burnout and patient safety among nurses (Li et al., 2024). Tawfik et al. (2019) conducted a meta-analysis of 123 studies that showed a quantitative positive relationship between health care professional burnout and poor quality of care, including patient safety. The studies included health care professionals working in outpatient, prehospital, emergency or inpatient settings. Similar results have been found in other health care settings. In a study of more than 200 long-term care organizations, results show that facilities with better patient outcomes (i.e. higher quality and safety of care, such as lower pressure ulcers and falls, higher patient satisfaction) were also those with better worker outcomes (i.e. fewer injuries, lower turnover, higher job satisfaction) (Boakye-Dankwa et al., 2017).

The positive relationship between patient safety and worker safety is partly explained by the similarity in factors that contribute to both outcomes. In line with the systems approach advocated in this book, we need to look at the work system factors that contribute to both patient safety and worker safety.

1.3.2 Patient Safety and Health Care Worker Safety Share Contributing Factors

Many studies have identified factors that contribute to both patient safety and health care worker safety. For instance, as mentioned above, high workload has a negative impact on both patient safety and various aspects of health care worker safety, such as fatigue and overall stress of ICU nurses (Gurses et al., 2009) and burnout of physicians and nurses (Aiken et al., 2023). Extended work duration and poorly designed work schedules contributed to both patient safety issues (e.g. medication errors and nosocomial infections) and worker injuries (e.g. needlestick injuries) in a study of more than 11,000 nurses (Olds & Clarke, 2010). Various characteristics of nursing work schedule, e.g. long work hours and lack of time away from work, were associated with adverse patient outcomes, such as postoperative PE and deep vein thrombosis and hemorrhage (Trinkoff et al., 2011a) and pneumonia deaths (Trinkoff et al., 2011b). In nursing jobs, high psychological and physical demands, which are known to affect worker health and safety (McCaughey et al., 2016), have also been shown to produce a negative impact on hospitalized patients, such as pneumonia and respiratory failure (Trinkoff et al., 2011a). More generally, negative characteristics of the physical, psychosocial and organizational work environment have been associated with negative outcomes for health care workers as well as suboptimal quality and safety of care (Aiken et al., 2023; Etchgaray & Thomas, 2014; Lundstrom et al., 2002; Wåhlin et al., 2019; Williams et al., 2012).

Besides specific job characteristics such as those identified above, a range of organizational factors have been found to influence both patient safety and health care worker safety. High organizational commitment to safety and collaboration in hospital units have been related to lower nurse injuries (i.e. needle-stick, splash, slip, trip or fall) and better patient safety (e.g. lower incidence of pressure ulcers) (Taylor et al., 2012). Patient safety culture and climate are strongly related to culture and climate for worker safety, and, therefore, are integral to influencing outcomes of patient and worker safety (Hesgrove et al., 2024). Psychological safety, or the perception related to the consequences of taking interpersonal risks (Edmondson, 1999), is an important feature of health care organizations to create conditions for improving both patient and worker safety (Maben et al., 2023). For instance, supportive leadership can enhance psychological safety and speaking up at work; therefore, allowing organizations to identify what does not work for both patients and health care workers. This positive organizational climate is also critical in safety improvement projects.

The relationship between patient safety and worker safety and the important role of common contributory factors have been emphasized by numerous professional organizations. For instance, in 2016, the American College of Occupational and Environmental Medicine issued a position statement about the linkage between patient safety and health care worker safety (Loeppke et al., 2017). They identified 12 categories of factors that affect the safety of both patients and workers, including physical factors (e.g. built environment, physical violence, patient handling), psychosocial factors (e.g. patient experience, psychological safety), organizational factors (e.g. communication, organizational culture, staffing) and biological and chemical factors (e.g. radiation, bloodborne pathogens).

Since patient safety and worker safety share contributing factors, the question is then how to link programs and approaches for simultaneously improving patient safety and worker safety. The 2012 Joint Commission report on improving patient safety and worker safety included an interesting subtitle: 'Opportunities for Synergy, Collaboration and Innovation' (The Joint Commission, 2012). This reinforces the idea that efforts at improving patient safety and those at improving worker safety should be at the minimum coordinated, if not integrated. This is also the perspective of the US Occupational Safety and Health Administration or OSHA (2024): 'Integrating (or at least aligning) patient and employee safety programs makes sense' (page 2). This fits with the SEIPS approach anchored within the HFE discipline, which aims at improving multiple outcomes for multiple stakeholders.

1.3.3 Patient Safety and Health Care Worker Safety May Conflict

As shown above, the evidence is clear that patient safety and health care worker safety are positively related to each other and can be affected by the same factors. But what happens in efforts aimed at improving either patient safety or worker safety? Are safety improvement activities designed and implemented so that both outcomes improve? Unfortunately, this may not always be the case. Some poorly designed patient safety interventions may actually create more work and increase workload, time pressure and clerical burden for frontline health care workers; this can lead to job dissatisfaction and distraction from providing high-quality, safe patient care (NASEM, 2019).

The process by which patient safety interventions are implemented is also important to consider. Poor implementation processes can create workflow disruptions and produce fear, stress and uncertainty among workers, which can negatively affect both workers and patients (Jones et al., 2024). When employees speak up about patient safety problems, what is the organizational response? Psychological safety is an important attribute of patient safety interventions (e.g. speak-up) that can also help to reduce or anticipate stress and anxiety among health care workers and enhance their overall well-being (Maben et al., 2023).

The tension between patient safety and worker safety is also felt by leaders and policymakers of health care organizations. In many countries, different components of the external environment (e.g. legislation, national reporting) are at play for patient safety and health care worker safety; this may lead to separate structures and processes within health care organizations, which add time and workload and may produce conflicts. In a series of studies in long-term care in Norway (Magerøy et al., 2023, 2024), leaders of nursing homes and policymakers clearly described the connection between patient safety and worker safety: 'They argued it was holistic work that could not be divided' (Magerøy et al., 2024, p. 08). They also talked about the potential conflicts and trade-offs between patient safety and health care worker safety. Some of these were due to the broad scope of handling two large objectives, such as patient safety and health care worker safety. This required a large responsibility with a lot of work, and trade-offs were often necessary when balancing resources and demands for addressing patient safety and worker safety. On the other hand, leaders of nursing homes talked about their holistic approach to change management; i.e. when a change was implemented to address one outcome, consequences for the other outcome were also addressed. This is a marker of the systems approach discussed in this book.

The goal should be to design and implement interventions that improve both patient safety and health care worker safety. This is the goal of HFE: designing work systems that produce positive outcomes for both patients and workers. This relies on a deep understanding of work system factors and knowledge of theories and models of job and organizational design, job stress, occupational health and safety sciences and HFE. This literature has largely contributed to our understanding of what makes for safe and healthy organizations for both patients and workers (Lim & Murphy, 1999; Sainfort et al., 2001; Sauter et al., 1996); this is what I call design principles or criteria that one can use to improve the design of work systems. This is further described in Chapter 3 on SEIPS and Chapter 4 on Human-Centered Design and SEIPS in action.

In this book, I discuss in detail the SEIPS approach with the goal of creating healthy work systems that produce safe and high-quality care for patients, while simultaneously providing safe and productive work systems for health care workers. As shown above, patient safety and health care worker safety share similar contributing factors; therefore, our focus should be on ensuring that the contributing factors are addressed in a human-centered design (HCD) process aimed at creating healthy work systems and healthy organizations (Lim & Murphy, 1999; Sainfort et al., 2001).

1.4 HEALTH CARE SAFETY

Health care safety is about the combination of patient safety and health care worker safety. As described in the previous sections, I use a broad conceptualization of health care safety with multiple dimensions of patient safety and health care worker safety. I also include the perspective of care partners when discussing health care safety.

1.4.1 Broadening Health Care Safety

My approach to health care safety is in line with suggestions by Vincent and Amalberti (2016), who recommend to go beyond 'traditional' aspects of patient safety and advocate for a broader vision of patient safety. They state that 'harm has been defined too narrowly' (page 3). In line with Safety-II proposed by Hollnagel (2014), they suggest that we need to look at what promotes or reinforces safety; this does not preclude the analysis of what does not work and adverse safety events. They also propose that patient safety issues should be broadened and go beyond what is traditionally viewed as the perimeter of patient safety, e.g. medication safety, HAIs and pressure ulcers. They recommend including other risks and harm to patients, such as malnutrition, dehydration and delirium that are relevant in particular for older adults (Long & Vincent, 2012). This broad approach of patient safety is aligned with the recommendation of Pronovost and colleagues who argue that harms tend to cluster and patients often experience multiple harms simultaneously (Pronovost & Bo-Linn, 2012) and that we need to include a broader range of harms, such as patient feeling of disrespect (Pronovost et al., 2017).

I broaden the patient safety view proposed by Vincent and Amalberti (2016) and others and include the perspective of health care workers; it is important to design work systems and care processes that not only promote safety for patients but also safety for health care workers. This is in line with the HFE discipline that aims at improving multiple outcomes, including performance, well-being, occupational health, and safety and quality of working life (International Ergonomics Association (IEA), 2000). Therefore, health care safety is about safety for both patients and workers.

We need to address safety objectives that are important to patients and health care workers. Landa-Avila et al. (2021) proposed an interesting approach to identify inter-connected objectives from the perspectives of patients and health care workers. Using a participatory outcome mapping approach, they found that patients and health care workers have many objectives in common, such as patient well-being and quality of life. The two groups have also unique objectives; for instance, patients mentioned personal resilience and happiness, and health care workers mentioned access to care and readmissions. In any improvement project aimed at health care safety, it is important to consider multiple inter-related safety outcomes for multiple stakeholders. This calls for an approach that considers multiple perspectives (e.g. patients, health care workers) in a participatory design process. Section 4.8 in Chapter 4 on Human-Centered Design and SEIPS in Action includes a discussion about how to integrate multiple perspectives in human-centered design.

Remember that when I talk about outcomes and objectives for patients, I also mean to include their care partners. See Section 1.4.3 for a discussion about the critical role of care partners in health care safety.

1.4.2 Health Care Safety As a 'Chronic' Goal

The HFE discipline emphasizes the goal of improving both performance and well-being (Dul et al., 2012; International Ergonomics Association (IEA), 2000). From an HFE perspective, we can then ask the following questions: do patient safety and worker safety result in convergent or conflicting outcomes? Do their contributing factors overlap, are they distinct or do they diverge? As discussed in Section 1.3.1, patient safety and worker safety may converge as similar work system characteristics may influence them, but they may also conflict. Patient safety and worker safety may not have the same contributing factors or efforts aimed at improving patient safety may have a negative impact on worker safety, or vice versa. In adopting the term 'health care safety', I emphasize the need to design work systems and processes that provide benefits for both patients and health care workers. I propose that human-centered design (HCD) can be a way of achieving these benefits. The challenging objective of health care safety requires long-term dedication and learning; time is essential. Therefore, in line with Woods (2006), I use the term safety as a 'chronic goal' that requires leadership, resources (including HFE knowledge and skills), time and sustained attention.

Woods (2006) distinguishes between acute and chronic goals of an organization. In health care, the Institute of Medicine (2001) defined six quality goals: patient care should be safe, effective, patient-centered, timely, efficient and equitable. Later on, the six goals evolved into three goals in the form of the Triple Aim (Berwick et al., 2008): improving care experience, improving population health and reducing costs of health care. The Triple Aim was then enlarged to the Quadruple Aim by adding the goal of caring for health care workers (Bodenheimer & Sinsky, 2014), and recently to the Quintuple Aim by adding health equity (Nundy et al., 2022). We are slowly going back to the original six goals of the 2021 IOM report with the addition of improving working conditions for health care workers; this additional focus on health care workers is a good and important addition that fits with the double objective of HFE and my approach to health care safety. I will now go back to Woods' description of acute and chronic goals in the context of the original IOM goals: the acute goals are effectiveness, timeliness and efficiency, and the chronic goals are safety, patient-centeredness and equity.

Woods (2006) describes inherent tensions and trade-offs between acute goals (i.e. effective, timely and efficient care) and chronic goals (i.e. safe, patient-centered and equitable care). Acute goals tend to have a short-term perspective, whereas chronic goals tend to evolve and improve over the long-term. As there is a tendency to emphasize short-term and more immediate objectives, acute goals often come first and sometimes at the expense of chronic goals. When acute goals are prioritized, chronic goals may actually 'lose'. For instance, creating new health care units or organizations may improve acute goals such as effective care for specific patient populations, while at the same time potentially increasing fragmentation of care and, therefore, potentially undermining patient-centered care. Woods argues that chronic goals should be prioritized over acute goals. This is why I choose to use the term safety in the umbrella concept of 'health care safety'. Of course, acute goals are important. Health care workers as well as patients often experience the negative consequences of inefficient and ineffective care processes. But the focus should be on long-term effort aimed at improving work systems and care processes for the benefit of both patients and workers. Therefore, we need to include multiple perspectives and confront them in a facilitated process such as human-centered design or HCD.

1.4.3 Safety for Patients and Care Partners

In this book, when I talk about safety for patients, I also mean to include their care partners. Patients interact with health care and are often not alone during those interactions, such as a child with their parents, a patient with a family member or an older adult with their spouse or a friend. Most literature uses the term 'caregivers' or 'informal caregivers' to describe those various roles. I use the term 'care partner' as suggested by the spouse of an older adult patient who interacted with the ED multiple times. The spouse participated in one of our projects and recommended that we use the term 'care partner' instead of 'caregiver'; I agree with her.

Care partners have unique roles and experience specific challenges. For instance, the *physical environmental* design of hospitals (e.g. space for care partners, layout of hospital room) may limit or facilitate the participation of care partners in their loved ones' hospitalization (Still et al., 2025). Care partners bear responsibilities for many activities and often experience high workload, especially when helping and supporting patients with chronic conditions; this can lead to caregiver burden (Adelman et al., 2014), stress and burnout (Gerain & Zech, 2019).

Care partners have a critical role in ensuring that their loved ones receive high-quality and safe care, in particular during vulnerable moments of the patient journey, such as care transitions between the hospital and home (Werner et al., 2019). Emerging HFE research on care partners analyzes the unique work systems of care partners (Holden, Schubert, & Mickelson, 2015; Look & Stone, 2018; Ponnala et al., 2020). For instance, Ponnala et al. (2020) described the caregiving activities of care partners of persons living with dementia; they also identified the networks and roles of individuals that care partners interact with. We need more research to understand the systemic aspects of care partners' activities and their role in health care safety (Holden et al., 2020; Werner, Ponnala, et al., 2020).

Improving health care safety means improvements for patients and their care partners, as well as the health care workers whom they interact with.

1.5 WHY IS IT DIFFICULT TO IMPROVE HEALTH CARE SAFETY?

As explained in Section 1.2, the COVID-19 pandemic has clearly shown how structural characteristics of health care contribute to negative outcomes of both patient safety and health care worker safety. It is also clear that a lot of progress is still needed to improve health care safety in a significant manner. There are many reasons for the lack of improvement in health care safety, including the limited use of systems approaches and insufficient involvement of multiple disciplines. Health care safety problems are too complex and often complicated and wicked to be addressed without knowledge and methods from multiple disciplines, in particular HFE and safety sciences.

1.5.1 Lack of Improvement in Health Care Safety

Are we making progress on patient safety? Editorials and commentaries published for several anniversaries of the publication of major reports in the United States and the United Kingdom have all indicated a lack of improvement in patient safety (Bates & Singh, 2018; Martin et al., 2023; Schiff & Shojania, 2022; Shojania & Thomas, 2013; Wachter, 2010). A 2023 report by the UK Imperial College on global patient safety stated: 'Many of the indicators show *very little improvement in performance over time*, often stretching back further than the data period under analysis in this report' (Illingworth et al., 2023). A 2024 report by the US Agency for Healthcare Research and Quality (AHRQ) shows improvements in several patient safety indicators, but major disparities in patient safety exist between care settings, demographic groups, regions, etc. (Agency for Healthcare Research and Quality (AHRQ), 2024). Even if we see some signs of patient safety improvement over time, the improvements are small and limited to specific domains (Shojania & Thomas, 2013) and health care-related harm still occurs. A retrospective study of 2,809 admissions to 11 hospitals in the state of Massachusetts during the 2018 calendar year shows that 1% of patients experienced a preventable adverse event with at least serious severity (Bates et al., 2023). In the UK National Health Service (NHS), organizational failures still occur, even after the systemic safety events that happened at the Mid Staffordshire NHS Foundation Trust and the subsequent numerous recommendations issued by a major public inquiry conducted in 2013. The so-called Francis report examined the failures that occurred in the main Trust in the Stafford region of the United Kingdom between 2005 and 2008 and proposed a wide range of recommendations at multiple levels of the health care system (Mid Staffordshire NHS Foundation Trust Public Inquiry, 2013). Ten years after the publication of the Francis report, Martin et al. (2023) indicated that many recommendations included in the report were not followed by changes at the policy or health care organization levels and that 'organizational catastrophes' still occur. Numerous reasons have been proposed for the lack of progress in patient safety, as well as health care worker safety.

Reasons for the lack of safety improvement include:

- the need to improve measurement of safety and develop measures for surveillance as well as evaluation of interventions (Martin et al., 2023; Shojania & Marang-van de Mheen, 2020; Shojania & Thomas, 2013),
- limited follow-up to recommendations and poor dissemination and implementation of recommendations (Martin et al., 2023),
- insufficient consideration for accountability and collaboration at the level of the entire health care sector (Dixon-Woods & Pronovost, 2016),
- lack of organizational leadership and commitment for safety (Dixon-Woods, 2010; McCulloch, 2024; Schiff & Shojania, 2022), and
- insufficient attention and effort to tackle cultural issues such as fear, blame and barriers to speaking-up (Martin et al., 2023; Schiff & Shojania, 2022).

Schiff and Shojania (2022) identified multiple challenges to improving patient safety, including challenges at the level of the health care sector (e.g. fiscal constraints, competing priorities), at the level of care settings (e.g. persistence of fear and blame, neglect of system interventions) and at the level of staff (e.g. lack of time and support, burnout). Interestingly, several of these challenges to improving patient safety are also related to a lack of improvement in health care worker safety. For instance, 'lack of time and support for both doing clinical jobs and improvement work' is relevant for both patient safety and health care worker safety. This calls for a systemic approach that can simultaneously address multiple dimensions of health care safety in a continuous cycle of learning and improvement.

When adverse events continue to occur even after patients and health care workers have experienced harm, one may ask whether the organizational structures involved in health care safety are adequate. In a commentary written 10 years after the publication of the Francis report on adverse events and organizational failures at the Mid Staffordshire NHS Foundation Trust in the United Kingdom, Martin et al. (2023) emphasized learning, i.e. 'gathering, collating, and acting on intelligence, quantitative and qualitative, formal and informal, leading and lagging' (page 1). It is important to not only learn from adverse events but also learn from safety improvement interventions and programs. Learning is also about acting on information gathered on events, near misses and interventions.

Improving health care safety is difficult; these are complex, complicated and/or wicked problems that unfold and evolve over time. It can be difficult to see 'cause' and 'effect' relationships; there may actually not be a 'cause' or 'causes,' but a web or network of factors that interact and over time lead to adverse events. In addition, health care work systems and care processes are dynamic and change constantly. Therefore, learning and action for improvement should be continuous.

1.5.2 Need for Systems Approaches to Health Care Safety

In previous sections, I have indicated a few problems with how we conceptualize health care safety, including the fragmented approach to analyzing and improving health care safety domains, especially as we know that they share contributing factors and that many safety domains are interdependent and inter-related. We tend to look at safety domains independently from each other, instead of using a systems approach to improve inter-related safety domains. While many experts have called for increased and significant integration of systems thinking and HFE for improving health care safety (Braithwaite et al., 2020; McCulloch, 2024; Schiff & Shojania, 2022), some also discussed problems with these approaches, such as their heterogeneity (Shojania & Thomas, 2013) and their inability to find the perfect solution (McCulloch, 2024). These so-called 'problems' may reflect a lack of alignment about scientific and conceptual views regarding safety and misunderstanding of systems thinking and HFE (Russ et al., 2013; Wears & Sutcliffe, 2020). It is interesting to note that systems approaches are often in quotation marks, for example, 'systems perspective' in patient safety (Shojania & Thomas, 2013) or the 'systems' approach to patient safety (Dixon-Woods, 2010). Systems thinking is present in many disciplines and can be confusing to understand; this is addressed in Chapter 2 on Systems Thinking in Health Care Safety.

Systems approaches are heterogeneous and typically address structural, organizational, social and cultural factors embedded in the way health care work is done. For instance, systems approaches may target teamwork, i.e. how coordination, communication and cooperation can be improved in teams, and the work systems in which teamwork occurs (Morey et al., 2002; Salas et al., 2008). This type of effort is difficult to implement and evaluate, especially with traditional health services research methods (Shojania & Thomas, 2013). However, that does not mean that we should give up on those efforts; instead, we need to be creative and develop new methods for designing, implementing and evaluating these systems approaches. In addition, we need to challenge commonly held assumptions regarding health care safety interventions.

In health care safety, much focus is on identifying effective safety interventions that could then be disseminated and implemented throughout health care. These are conceptualized as solutions with the illusion that, if they work in one context, care setting or health care organization, then we should be able to implement them effectively in other environments. This ignores our deep knowledge about the organizational, social and cultural aspects of health care changes and innovations

(Dixon-Woods, 2010; Greenhalgh et al., 2004). What works in one health care organization or care setting may not work in another. Adaptations may be necessary to integrate an intervention into a different work system. Attention should be paid to the organization receiving the innovation or change, including staff attitudes and expectations regarding the change and organizational commitment to safety. A narrative review of four projects on patient safety in UK hospitals identified several socio-organizational dimensions that can limit safety improvement (Dixon-Woods, 2010): organizational commitment to improving safety, communication and teamwork, tension between rules and workarounds, structures of authority and accountability and structure and culture for reporting safety incidents. If these dimensions are not systematically addressed and considered in a project or program for safety improvement, it is unlikely that efforts will produce positive outcomes. Therefore, we need to consider the local work system (where patients and health care workers interact), the organization(s) surrounding the work system and the external environment; this multi-level approach is key to systems thinking. HFE systems approaches, such as SEIPS, play an important role in this effort; but of course, other disciplines, models, theories and approaches need to complement them.

Multidisciplinary collaboration should be encouraged and supported to tackle the wicked problems of health care safety. A study on safety of fetal monitoring shows how limited and possibly misdirected approaches have hindered progress in preventing harm to mothers and newborns; the authors proposed to integrate systems approaches such as SEIPS with knowledge from other disciplines (Lamé et al., 2024). Researchers conducted observations and interviews in three maternity units in the United Kingdom to analyze daily work conducted from the time when the decision had been made to begin electronic fetal monitoring (EFM) during labor up to the point of escalation. Results show that EFM is a collaborative process that is affected by multiple interacting work system barriers. For instance, communicating concerns about potential fetal deterioration was hampered by challenges in interpreting the EFM signal and variable workflows regarding who the midwives should contact [*interaction between the person, task and organization elements of the work system*]. The authors' conclusion applies to many other health care safety problems: 'A broad set of influences on safety and their interdependencies must be considered in creating the conditions for improvement' (page 254). This calls for approaches that combine HFE systems models, such as SEIPS and knowledge from other disciplines, such as social and organizational sciences.

1.5.3 Genuine Multidisciplinary Collaboration

Multiple reports and organizations around the world have advocated for partnerships across various disciplines to improve health care safety, such as those described in Chapter 2 between engineering and health care. However, these partnerships have been limited. In a critical analysis of the patient safety movement, Wears and Sutcliffe (2020) describe a movement that is misled and addresses safety in a superficial or simplistic manner, while not really listening to and involving other disciplines. We can also see similar trends in efforts aimed at addressing health care worker safety, such as burnout and well-being. Whereas there is a rich and deep literature that addresses worker safety, burnout, quality of working life and well-being, this knowledge and experts in the disciplines of HFE, job and organizational design, and job stress are not effectively integrated in safety improvement projects. This unfortunately leads to organizational efforts that target the individual, whereas we need to redesign work systems and care processes. As 'the task [of improving safety] is more difficult and complex than initially thought' (Wears & Sutcliffe, 2020, p. 4), we need to (re-)commit to genuine collaboration across disciplines in order to significantly and continuously improve health care safety.

We need to change the way we think about and conceptualize health care safety and the way we act to improve safety (Pronovost et al., 2017). This requires a significant focus to understand the complex networks in which health care work occurs and safety is created or deteriorates (Dixon-Woods, 2010). This begs the question of whether the right disciplines (including HFE) are involved in health care safety (Wears & Sutcliffe, 2020). In a chapter in the book edited by Sue Bogner (1994) on 'Human Error in Medicine', John Senders (1994) stated: 'Human error in medicine, and the adverse events that may follow, are problems of psychology and engineering, not of medicine' (page 159). Therefore, the necessary change in mental models about safety requires the genuine involvement of multiple disciplines, in particular those interested in understanding how people work (e.g. HFE, design sciences, safety sciences, work and organizational psychology, engineering psychology). HFE, in particular, can play an important role in health care safety.

1.6 WHAT IS THE ROLE OF HFE IN HEALTH CARE SAFETY?

Everything in health care is about people: patients and care partners interact with health care organizations and their workers. Therefore, in efforts aimed at improving health care safety, we need an approach that gives serious consideration to people. HFE is a discipline aimed at creating human-centered systems and processes with the goals of supporting the performance and well-being of all those involved. HFE is based on humanistic values that promote empathy and benefits for all involved. Therefore, HFE should play a critical role in health care safety improvement.

Various reports around the world have called for HFE in health care safety. In the United States, the National Academies have issued multiple reports on health care safety, starting with the 2000 report on 'To Err is Human: Building a Safer System' (Kohn et al., 2000). This report includes a chapter on 'Why do errors happen' that draws heavily on HFE and safety sciences. It emphasizes the need to look at latent factors that contribute to errors and to focus safety efforts on system design. The 2001 report on 'Crossing the Quality Chasm: A New Health System for the 21st Century' (Institute of Medicine, 2001) calls for HFE knowledge to be 'applied in designing tasks, processes, equipment, rules, and environments' (page 78). The 2004 report on 'Keeping Patients Safe: Transforming the Work Environment of Nurses' calls for attention to the work system of nurses who are at the forefront of patient care and have a major role in patient safety (Institute of Medicine, 2004). It includes a chapter on work design that uses the work system model (Carayon, 2009; Smith & Carayon-Sainfort, 1989) and proposes five questions on the work system elements (see Table 4.1 in Chapter 4 on SEIPS). The 2012 report focuses on 'Health IT and

Patient Safety' (Institute of Medicine, 2012) and discusses specific HFE issues and methods, including usability and user-centered design. The report proposes a model of the sociotechnical system that has many similarities to the work system model; it includes the following elements: *people, technology, process, organization and external environment.* The 2018 report on 'Crossing the Global Quality Chasm' (NASEM, 2018) proposes principles for redesigning health care using systems thinking, HFE and human-centered design (HCD). This report also calls attention to the patient journey and the integration and coordination of health services over one's entire lifetime; this is the focus of the SEIPS 3.0 model of patient journey (Carayon, Wooldridge, et al., 2020). The 2019 report on 'Taking Action Against Clinician Burnout: A Systems Approach to Professional Well-Being' (NASEM, 2019) recommends the use of HFE and HCD to address multiple contributors to clinician burnout, such as workload related to standards and regulations, and poorly designed health information technologies. The conceptual framework of clinician burnout in this 2019 report integrates the work system model (Carayon, 2009; Smith & Carayon-Sainfort, 1989).

The Global Patient Safety Action Plan of WHO (2021) has five strategic objectives, including objective 2 'to build high-reliability health systems and health organizations that protect patients daily from harm'. HFE is a major component of objective 2. A 2024 report by WHO shows that significant effort is necessary to meet objective 2 (World Health Organization (WHO), 2024):

> "Although the significance of human factors in health care is being increasingly recognized at the global level, only around a quarter of countries have started to apply human factors principles in patient safety interventions in clinical practice, use of medical devices, information technology (IT) solutions, and service delivery processes."
>
> (page 77)

Considerable effort remains to address the recurring call by international organizations for HFE involvement in redesigning health care work systems and processes for better health care safety around the world. Hopefully, this book will help to further disseminate ideas, concepts and methods about HFE and SEIPS.

2 Systems Thinking in Health Care Safety

In recent years, systems thinking has received increased attention as a means of understanding, addressing and solving health care problems such as patient safety and quality of care. This is reflected in many reports published by the US National Academies. For instance, the 2019 report by the National Academies includes 'systems approach' in its title ['Taking Action Against Clinician Burnout: A Systems Approach to Professional Well-Being'] and uses HFE systems models, including SEIPS (NASEM, 2019). Many international institutions have also emphasized systems thinking in relation to patient safety. For instance, the WHO Global Patient Safety Action Plan incorporates a systems approach aimed at improving safety at multiple levels of action (including HFE) (World Health Organization (WHO), 2021). However, some applications of 'systems thinking' have a limited understanding of systems principles and approaches. A so-called 'systems approach' may be applied, but may not reflect true systems thinking; therefore, leading to an 'illusion' of applying systems thinking or, in some cases, a complete misunderstanding.

CONFUSION AND MISUNDERSTANDING ABOUT SYSTEMS THINKING

A series of papers published in BMJ Quality & Safety in 2014 and 2015 (Dekker & Leveson, 2014, 2015; Levitt, 2014a) in reaction to a commentary by Levitt (2014b) illustrates the misunderstanding of systems thinking and the systems approach in patient safety. In a commentary published in the *Los Angeles Times*, Levitt (2014b) criticized the 'widespread embrace of systemized medicine in hospitals in the country' and argued that the 'systems approach' adopted by health care did not improve patient safety. He described the 'systems approach' as a way of 'standardizing routine aspects of patient care', for example, with the use of checklists. He called for a refocus of patient safety on incompetent physicians. Dekker and Leveson (2015) explained the misunderstanding regarding the systems approach that Levitt equated to standardization, checklists and other elements. They actually called for further adoption of the systems approach 'in order to improve the opportunity to do the right thing'. In response to the commentary of Dekker and Leveson (2014), Levitt (2014a) continued to misrepresent the systems approach as a set of 'standardization techniques to correct defects' and insisted on the need to identify incompetent physicians or so-called 'bad apples'. In their final response, Dekker and

DOI: 10.1201/9781003480310-2

Leveson (2014) explained that the so-called 'bad apples' can actually be considered a system problem. The way health care is organized and designed influences behaviors and attitudes. Therefore, the question becomes: how can we design a work system to identify, prevent and reduce problematic or toxic behaviors and attitudes? This exchange shows how the systems approach to health care safety can be misunderstood; therefore, we need to be clear about terminology and concepts.

The term 'system' is used to mean different things, sometimes not in line with systems thinking and its core foundational concepts; see Section 2.4 later in this chapter. Additional confusion is created as different terms are used to talk about systems thinking: e.g. systems approach, systems science, systems change, etc. In this chapter, I clarify the domain of systems thinking, in particular its application to health care safety with the discussion of complex systems. I then focus on the concept of work system and discuss how the HFE discipline incorporates systems thinking.

2.1 SYSTEMS THINKING AND SYSTEMS APPROACHES TO HEALTH CARE IMPROVEMENT

Systems thinking can be broadly defined as a way to think in systems, which is the basis for problem analysis, solution design and evaluation and continuous improvement (Adam & de Savigny, 2012; Cabrera & Cabrera, 2019). Systems thinking is found in various fields and has been developed in multiple disciplines, such as mathematics, engineering, life sciences and social and behavioral sciences (Cabrera & Cabrera, 2019; Monat & Ganon, 2015). Systems thinking often refers to a range of methodologies and tools that describe, model, analyze and help to improve complex systems such as those found in health and health care (Carey et al., 2015). Systems thinking is very diverse; its core concepts can be summarized as follows (see Table 2.1):

1. holistic orientation to systemic problems,
2. recognition of complex network structures with multiple dynamic interactions among system elements and levels and
3. multiple perspectives to address a problem or event.

2.1.1 Waves of Systems Thinking

Systems thinking has received increasing attention, in particular in health care. However, confusion exists about what systems thinking is, given the wide range of disciplines and approaches. Cabrera and Cabrera (2019) described four waves of systems thinking.

TABLE 2.1
Core Concepts of Systems Thinking

Core Concepts of Systems Thinking	Description
Holistic orientation to systemic problems	To solve problems, we need to focus on the system as a whole, not just the elements or parts of the system.
Recognition of complex network structures with multiple dynamic interactions among system elements and levels	We need to appreciate that behaviors and outcomes emerge because of dynamic system interactions. Be suspicious of simple linear causal mechanisms.
Multiple perspectives to address a problem or event	Various people and stakeholders have different mental models, and therefore, different ways of understanding the problem at stake and designing possible solutions.

2.1.1.1 First Wave of Systems Thinking: Quantitative Methodologies

The first wave focused on 'hard' systems and developed primarily quantitative methodologies, such as those found in the operations research discipline. For instance, the system dynamics (SD) approach aims to model social, ecological or organizational systems and their dynamic interactions and evolution; for information on the past, present and future of SD, see papers by Forrester (2007a, 2007b). SD produces diagrams showing causal relationships between system elements and highlighting feedback loops that reveal system behavior over time. SD uses computer-based simulation to implement a series of mathematical equations that quantify relationships and feedback loops (e.g. stock and flow diagram). A review of literature of 63 studies describes various applications of SD to health care safety (Ibrahim Shire et al., 2018). These studies have used SD to analyze safety events (retrospective analysis) as well as to assess the potential effectiveness of solutions or policies (prospective analysis).

2.1.1.2 Second Wave of Systems Thinking: Quantitative and Qualitative Methodologies

The second wave of systems thinking is associated with researchers such as Ackoff (1971) and Checkland (1999), and focuses on methods to better appreciate systems, in particular those that include people. In this wave of systems thinking, both quantitative and qualitative methodologies were developed. For instance, soft systems methodology (SSM) proposes an approach to system improvement with four stages: (1) finding out about the problem, including its cultural and political dimensions; (2) coming up with activity models; (3) discussing the problem, using the models; and (4) implementing a change in the system to bring about improvement (Checkland, 1999). In a scoping review of the application of SSM in health care, Augustsson et al. (2020) show the various problems addressed by 49 studies, such as improving care for children with serious emotional disturbance or designing information technologies to improve communication and information sharing, e.g. between various facilities involved in rehabilitation care for older adults. An application of SSM by Markopoulos et al. (2023) aimed to develop a new model of intensive care unit (ICU) that incorporates various advanced information technologies. The goal of the new

ICU model was to enhance communication within the ICU and between the ICU and other hospital units and improve engagement of family members, which can produce benefits for health care staff, patients and care partners.

2.1.1.3 Third Wave of Systems Thinking: Combining Quantitative and Qualitative Methodologies

The third wave of systems thinking is called 'critical systems thinking'. It aims at countering the separatism that emerged from the first two waves, calls for combining multiple approaches and methodologies, and advocates stakeholder participation. An example of this is the combination of the quantitative and qualitative methods of SD with participatory ergonomics in designing efficient and safe hospital medication dispensing systems (Ibrahim Shire et al., 2019; Ibrahim Shire et al., 2020). An SD causal loop diagram was developed to show the negative impact of work pressure on dispensing errors and its potential positive impact on productivity. This was complemented by an SD quantitative simulation model to assess the impact of various work system characteristics (i.e. pharmacy staffing level, interruptions and fatigue) on medication dispensing errors. The causal diagram and the simulation model were developed and used in a participatory manner with a group of 21 health care professionals, including pharmacists, managers and trainees (Ibrahim Shire et al., 2020). The combination of SD tools and participatory ergonomics produced an environment for team learning and improved decision-making as trade-offs were more clearly understood. This study fits with approaches recommended in this book as it combines multiple models, concepts and disciplines.

2.1.1.4 Fourth Wave of Systems Thinking: Skills in Systems Thinking

The fourth wave of systems thinking emphasizes skills about systems thinking: how does one think in systems or see systems (Oshry, 2007)? In health care and patient safety, this has led to various models and approaches, such as the design of 'thinking systems' for care coordination of cancer patients (Randhawa et al., 2019) and the Systems Thinking for Everyday Work (STEW) approach (Fell et al., 2024; McNab et al., 2020), which is embedded in the HFE discipline and is described below.

2.1.2 Systems Thinking for Everyday Work (STEW)

The STEW approach is particularly relevant as it focuses on the actual work performed in health care organizations and aims at improving patient safety and care quality. The STEW principles were developed in a participatory co-design process with experts in systems thinking and safety, health care professionals working in various roles in primary care (e.g. physician, nurse, manager, pharmacist) and patient safety leaders (McNab et al., 2020). See Table 2.2 for a description of the STEW principles. The STEW principles are available as a set of cue cards on the NHS Scotland website: https://learn.nes.nhs.scot/6027.

McNab et al. (2020) described a case study of the application of the STEW principles in a primary care facility that hired a pharmacist. The case study shows how the STEW principles help to describe the impact of the pharmacist's appointment

TABLE 2.2
Systems Thinking for Everyday Work (STEW) Principles (McNab et al., 2020)

STEW Principles	Description
Foundation concept	• Apply a systems approach to health care improvement. • Agree on the boundaries of the system to improve. • Define the purpose of the system.
Seek multiple perspectives	• Appreciate that people in different positions and roles at all levels of the organization are experts in their work. • Explore multiple perspectives and views on the problem and potential solutions.
Consider work conditions	• Work conditions include demands, capacity, resources and constraints. • Examine how the interactions between work conditions influence the work done by people, and, therefore, their performance and well-being.
Analyze interactions and workflow	• Interactions between system elements are complex and dynamic and affect both performance and well-being. • When making a change to part of a system, be mindful of the impact of the change on the rest of the system. • Explore the flow of work (e.g. patient flow, information flow) to identify bottlenecks and blockages.
Understand why decisions make sense at the time	• Be wary of hindsight bias when examining events and past decisions. • Move beyond blaming 'human error' and promote a just culture.
Explore performance variability	• Work conditions change, and people adapt to these changes in order to achieve high-quality, safe care. • Consider how adjustments, such as trade-offs and workarounds, contribute to both successful and unsuccessful outcomes. • Understand the difference between work-as-done (WAD) and work-as-imagined (WAI).

on the work and processes within the primary care facility. This analysis led to several changes to improve performance and patient safety, such as ensuring that pharmacists had access to the hospital's medication prescription information system and organizing regular mentoring sessions between the pharmacist and the primary care physicians. The STEW card set is a practical tool that can help health care staff and leaders understand systems thinking and its application to patient safety (Fell et al., 2024).

2.2 SYSTEMS THINKING IN PRACTICE: ENGINEERING CONTRIBUTION TO HEALTH CARE IMPROVEMENT

The different waves of systems thinking described above have often involved expertise from engineering, in particular systems engineering (e.g. SD or system dynamics (Forrester, 2007b; Ibrahim Shire et al., 2018)). Interdisciplinary approaches to health care improvement have emerged over the years and have strongly encouraged collaboration between engineering and health care. In the United States, the National

Academies' report on 'Building a Better Delivery System. A New Engineering/ Health Care Partnership' (Reid et al., 2005) emphasizes the contribution of a systems approach based on engineering to address the six quality aims: safety, effectiveness, timeliness, patient-centeredness, efficiency and equity. The systems approach proposed in the report highlights levels of the system where engineering tools and methods can be applied:

- level of the individual patient,
- level of the care team that is composed of health care workers, patients and care partners,
- level of the health care organization, e.g. hospital, clinic and nursing home and
- level of the external environment, including regulations and payment systems.

The report organizes systems engineering tools into three categories: tools for design, tools for analysis and tools for control, which can be applied to address health care problems at different system levels. For instance, HFE provides a set of tools and methods to improve, e.g. the design of instructions for patients (level of individual patient), coordination and communication among care team members (level of care team), packaging of medications (levels of organization and environment) and medical education (level of the external environment).

In the United Kingdom, the Royal Academy of Engineering collaborated with the Royal College of Physicians and other professional and governmental organizations to propose a systems approach to address health care problems (Clarkson et al., 2017). Their systems approach includes four core elements or principles:

- *People* are at the center of systems as they ultimately contribute to success.
- *Systems* are composed of interconnected technical and social elements that produce emergent behaviors and actions.
- *Design* processes help to identify the right problem to solve, to explore potential solutions and to select and refine the best solution.
- *Risk* assessment is key to identity potential or actual hazards in order to manage and implement necessary changes.

Systematic reviews show that engineering-based systems approaches to health care intervention result in improvement for both patients and services (Komashie et al., 2021); this is true, in particular, for HFE-based systems approaches (Waterson, 2009; Xie & Carayon, 2015). However, confusion remains about what a systems approach means in practice (Waterson, 2009). While many claim to adopt a systems approach to patient safety, they fail 'to take a holistic, systems approach to the issue' (Wears & Sutcliffe, 2020, p. 175). This book describes a specific systems approach to health care safety by proposing a conceptual framework (SEIPS) and a method (human-centered design or HCD) to address complex safety problems. We need to address the confusion about systems thinking; this is what I do in the following sections in describing complex systems. I then conclude the chapter by describing the systems approach advocated by the discipline of HFE.

2.3 COMPLEX SYSTEMS IN HEALTH CARE

Systems thinking is often described as the opposite of linear thinking; interactions and relationships among system elements are more important than the elements themselves. From a systems thinking perspective, systems are complex and dynamic, and have emergent properties that result from interactions (Monat & Ganon, 2015). According to Greenhalgh and Papoutsi (2018), 'Complex systems have fuzzy boundaries; their interacting agents operate on the basis of internal rules that cannot always be predicted; and they adapt, interact and co-evolve with other systems' (page 2).

Several authors have described conceptualizations of complexity and the evolution of the concept of complexity over time (Hollnagel, 2012a; Walker et al., 2010). Complexity can be described by its multiple dimensions or attributes, such as dynamism, uncertainty and large problem spaces (Vicente, 1999; Walker et al., 2010). Health care possesses many of the complexity attributes; e.g. health care includes many different people, organizations and communities that can be either tightly or loosely connected and interact in dynamic and uncertain environments (Carayon, 2006). Describing the complexity attributes of health care is not sufficient; we need to go beyond the attribute view of complexity and identify implications of coping with and managing complexity to improve health care safety (Hollnagel, 2012a). Given the complexity in health care, how can we design and implement interventions with the goal of improving safety? One approach is to rely on principles derived from complexity theory, such as self-organization and sensemaking (Lanham et al., 2013).

2.3.1 Health Care as Complex Adaptive Systems

Several authors have conceptualized health care as complex adaptive systems (Braithwaite et al., 2017; Greenhalgh, 2018; Plsek & Greenhalgh, 2001; Rouse, 2008). They described five characteristics of complex adaptive systems and their implications for improving health care (see Table 2.3).

1. Complex adaptive systems have fuzzy boundaries. They are connected to other systems, are embedded in larger systems, and are composed of subsystems. Therefore, it is not easy to define the boundaries of the system to be analyzed and improved. Sometimes, the solution may lie outside of the system under study. In any health care safety improvement project, it is important to define the system and its boundaries; it is also important to be flexible and to understand the 'external environment' and the connected systems (system of systems) during the analysis and design process as solutions may be found outside of the boundaries defined initially. In addition, as systems of systems have no organizational structure common to the inter-connected systems, collaboration and action may be challenging when addressing health care safety problems that lie at the interface of care settings, organizations or organizational units (Siemieniuch & Sinclair, 2014).
2. The 'agents' in the system (e.g. patients, health care professionals, leaders of health care organizations) have mental models, which influence their perceptions, behaviors, actions, reactions, etc. In addition, the agents are adaptive: they react to the conditions of the system and act on the system. It is important to give autonomy to agents as they know the work at the 'sharp end' and can adapt to changing conditions and unpredictable events; this is the principle of 'deference to expertise' in the High Reliability Organization (HRO) Theory (Weick & Sutcliffe, 2001).

3. Tension and paradox are inherent in complex systems. We need to get used to this and accept that not all tension needs to be resolved; it may not be possible anyway. For instance, tension and paradox exist in the implementation of evidence-based health care in the form of guidelines, i.e. "systematically developed statements to assist practitioner decisions about appropriate health care for specific clinical circumstances" (Greenhalgh, 2014, p. 135). Guidelines may have ambiguity that cannot be easily resolved, such as ambiguity regarding who is responsible for implementing the actions recommended in a guideline or the misfit between what is recommended by a guideline and the constraints in the actual work system (Gurses et al., 2008). As tension may occur because of various viewpoints and objectives, multiple perspectives need to be considered in designing and implementing health care safety interventions.
4. Complex adaptive systems are dynamic as the agents or people and other system elements interact, adapt and change over time. Systems thinking is about interactions among elements or components of a system; system interactions may lead to emergent behavior or outcomes that cannot be predicted. It is important to expect surprises and be creative and adaptive. This requires processes that allow for continuous design and improvement and sensemaking when implementing health care safety interventions.
5. Even though complex adaptive systems are uncertain, dynamic and often unpredictable, we can manage and cope with complexity. Over time, complex adaptive systems develop patterns of behavior, local rules and routines. These patterns and structures influence actions, reactions and interactions of agents. They can be analyzed, described and leveraged to produce systems change (Woods, 2000).

TABLE 2.3
Implications of Complex Adaptive Systems (CAS) for Health Care Safety

Characteristics of CAS	Implications for Health Care Safety
CAS have fuzzy boundaries, are connected to other systems and are nested.	We need to be flexible in defining boundaries of the system that is the focus of the health care safety intervention. Collaboration is necessary at the interface of systems.
People have mental models that influence activities. They adapt to and adapt the system and its changing and sometimes unpredictable conditions.	People at the 'sharp end' (e.g. patient, health care worker) need to have some autonomy and their expertise should be recognized and respected [HRO principle of 'deference to expertise'].
Tension and paradox are inherent in CAS. We need to understand this and deal with potential conflicts and ambiguity.	Multiple perspectives should be involved in designing and implementing health care safety interventions.
CAS are dynamic as people and other system elements change. Emergent behaviors or outcomes occur as system interactions develop and change.	Health care organizations need to set up structures and processes for learning and for continuous improvement of health care safety.
CAS exhibit patterns and routines even though they are uncertain and dynamic.	Systems analysis can help to describe and leverage patterns and routines with the goal of managing and coping with complexity in designing health care safety interventions.

Note: HRO = High Reliability Organization (Weick & Sutcliffe, 2001).

2.3.2 Interactions and Emergence

Interaction is often thought of as interaction between people: people interact to share information, communicate or coordinate actions, make decisions and socialize. These social interactions are important for performance, safety and well-being. In systems thinking, the term 'interaction' has a broader meaning. System interactions are about interactions between system elements, between the system and its external environment, between multiple systems, between outcomes and between systems, their elements and outcomes. These interactions are dynamic: as systems and their elements and outcomes change over time, their relationships also change over time. This introduces uncertainty and unpredictability; this is how emergence comes about.

In an essay on emergence and safety, Lintern and Kugler (2022) describe emergent states as 'generated by qualitatively transformative interactions between system parts and between system parts and environmental factors' (page 654). We can do a lot when designing health care work systems and processes to promote safety. For instance, a proactive risk assessment (PRA) can identify potential vulnerabilities and hazards of a potential solution; this information can then be used to improve the solution (Carayon et al., 2011; Hoonakker et al., 2019). However, this does not ensure that issues will not arise once the solution is implemented; this is the meaning of emergence. System interactions occur that cannot be all predicted nor anticipated; these emergent states result from system interactions. We may not exactly anticipate how these emergent states will be, but we should anticipate that they will occur.

When implementing an intervention in complex health care systems, we should expect surprises. We do not know what the 'surprises' will be, but we need to be ready to find them, understand them and respond to them as quickly as necessary. This calls for robust and organized processes for organizational learning and continuous improvement (Hundt, 2012; Hundt et al., 2012). Further discussion of system interactions and emergence can be found in Chapter 3 on SEIPS.

2.4 WHAT IS A SYSTEM?

In health care, the term 'system' is used in many different ways, which can create confusion and misunderstandings. When the term 'system' is used, are we really talking about 'systems' as defined by systems thinking and systems approaches? Not always! For instance, health information systems represent technologies that process and distribute patient-related information. Health information systems or health information technologies are only one element of a broader work system. The term 'system' is often used to describe a 'health system', i.e. a group of health care organizations such as a hospital and primary care clinics. The human body is composed of multiple 'organ systems' such as the nervous system and the skeletal system. These various uses of the term 'system' in the health care and clinical literatures do not do justice to the rich literature on systems thinking and the multiple systems approaches developed by various disciplines and described above. This, therefore, contributes to the confusion about the term system. Even though the term 'system' can be confusing, I will persist and use it to describe one particular type of system, i.e. the work system.

BOX – WORK SYSTEM OR SOCIOTECHNICAL SYSTEM

In this book, I focus on work systems or sociotechnical systems that involve *people* performing *tasks* using *tools and technologies* in a *physical environment* and under *organizational conditions*. The work or sociotechnical systems have multiple interacting levels. In health care, we are often interested in systems of systems as people transition between different care organizations and settings during their patient and health journey.

2.4.1 Definition of a System

A system is typically defined as 'a complex of interacting elements' that together produce outcomes (Bertalanffy, 1950, p. 143). A system is more than the sum of its elements; interactions among elements are key when talking about systems. One important distinction is made between closed and open systems. Closed systems do not exchange information, materials or other resources with the external environment, whereas open systems are open to change and influence with the external environment (Bertalanffy, 1950, 1969). Katz and Kahn (1966) describe organizations as open systems with input, throughput, output, feedback and interactions with the external environment. In health care, we are dealing with open systems that interact with a wide external environment.

A system has a purpose or a goal, such as diagnosing and treating injured or sick people. However, as indicated by Siemieniuch and Sinclair (2015), 'this purpose may depend on the perception of the observer' (page 856), the participant or the stakeholder. Different people involved in a work system or care process may not have the same objective. Therefore, understanding the multiple perspectives involved in a care process is critical for improving it; this is further discussed in Chapter 4.

2.4.2 Work or Sociotechnical System

In this book, I use the term 'system' accompanied by 'work' (i.e. 'work system') or 'sociotechnical' (i.e. 'sociotechnical system'). My perspective is to examine the way work is actually performed by, for instance, physicians, nurses, patients and care partners, as well as teams composed of various health care workers and patients. The 'work system' has characteristics that help, hinder, constrain, facilitate or support activities performed by individuals in various environments or contexts. In this book, a work system and a sociotechnical system are used interchangeably.

In health care, we are concerned with systems that include *people* (e.g. physician, nurse, patient) performing *tasks* using *tools and technologies* in a *physical, social and organizational environment*. This type of system has been called 'sociotechnical system', a concept developed by researchers and practitioners at the Tavistock Institute in the United Kingdom (Emery & Trist, 1965; Trist, 1981; Trist & Bamforth, 1951). A landmark study of coal mining showed how a technological change (i.e. from short-wall to long-wall method of coal getting) had a major impact on the social organization, teamwork and collaboration across the temporal, physical and production stages of the coal mining process (Trist & Bamforth, 1951). This led to the concept of sociotechnical systems with 'the goal of integrating the social requirements of people doing the work with the technical requirements needed to keep the work systems viable with regard to their environments' (Fox, 1995, p. 92). The technology, the people and the rest of the work system are interdependent.

2.4.3 Sociotechnical Systems (STS) Theory

As the STS Theory espouses humanistic values, it is concerned with creating sociotechnical systems that produce a high quality of working life (Trist, 1981); this is also our goal with SEIPS.

The STS Theory defines three levels of system analysis (Trist, 1981):

- primary work systems,
- whole organizations and
- macrosocial systems such as communities or industrial sectors.

At the level of primary work systems, work should be designed to foster participation and autonomy and have characteristics that satisfy psychological, social and other needs of workers and produce a high quality of working life (e.g. job satisfaction, motivation) while meeting performance requirements and expectations of organizations. This has been further developed in various models and theories of job design (Davis, 1966; Davis & Wacker, 1987; Hackman & Oldham, 1976) and models of teamwork (e.g. autonomous work groups) (Cummings, 1978; Pasmore, 1988). At the level of whole organizations, various design principles have been proposed, such as joint optimization of social and technical systems, participation, few organizational layers or levels and self-organization (Davis & Wacker, 1987). The STS Theory is also concerned with systems larger than single organizations or companies, such as a network of companies and communities. This has led to widening the domains of application of STS Theory, including crime prevention (Mumford, 2003) and managing events with large crowds (Davis et al., 2014).

The STS Theory has produced multiple methods for analyzing sociotechnical systems at multiple levels (Pasmore, 1988), such as variance analysis to analyze care processes (Hamilton-Escoto et al., 2004) and the search conference to facilitate collaboration and cooperation of large distributed groups dealing with social problems (Emery & Purser, 1996).

Another important contribution of the STS Theory is the description of sociotechnical design principles. Albert Cherns proposed a first set of principles in 1976 and an updated version in 1987 (Cherns, 1976, 1987). Chris Clegg extended these principles in 2000 and categorized them into: meta-, content- and process-principles (Clegg, 2000). The meta-principles represent a worldview and values for sociotechnical design. The content-principles are about what the design should be. The process-principles are related to the how-to of the design process, which is itself a

TABLE 2.4
Principles of Sociotechnical Design

Meta-Principles

1. Sociotechnical design is systemic, extended over time and involves multiple individuals.
2. The sociotechnical system should meet the needs of patients and health care workers.
3. Sociotechnical design involves making choices.
4. There is no one best way of designing a sociotechnical system.

Content-Principles

5. Core processes should be integrated and coordinated.
6. Jobs should be well-designed, e.g. meaning in work, opportunity for learning and adequate demands.
7. All elements of the sociotechnical systems should be congruent and fit to support performance, safety and well-being.
8. The means of working should be flexibly specified (i.e. minimal criteria specification), and problems should be controlled at the source.

Process-Principles

9. People should participate in the process of designing their sociotechnical systems.
10. Multiple perspectives, skills, experience and expertise should be involved in the design process.
11. Resources and support are necessary in the design process.
12. Sociotechnical system design reveals political issues and potential conflicts that should be managed.

Adapted from Cherns (1976, 1987), Clegg (2000) and Greenhalgh (2018).

sociotechnical system. Cherns and Clegg insisted that the principles are interrelated even though they are presented separately. Table 2.4 shows a combined list of principles of sociotechnical design based on the principles of Cherns (1976, 1987) and Clegg (2000) and the revised list of Greenhalgh (2018).

The term 'sociotechnical' is often used to characterize a system that has technical and social elements or characteristics; sometimes without reference to the STS Theory. This is not the case with SEIPS as I use various aspects of STS Theory, including its systems approach, human-centered design, humanistic values that are the basis of job design criteria, the focus on quality of working life and community learning.

2.4.4 System Levels

Systems are embedded in larger systems and are composed of sub-systems; this is what we mean by system levels. System level is a key concept in systems thinking as shown in Section 2.3 on complex systems and in the discussion on the STS Theory in Section 2.4.3. The STS Theory defines three levels of system: work system, organization and external environment (e.g. communities, industrial sector).

Many HFE and safety models incorporate multiple system levels. For instance, Carayon, Hancock et al. (2015) used the three levels of STS Theory to propose a model of workplace safety represented by three concentric circles (see Figure 2.1). The first concentric circle is the work system as proposed by Smith and Carayon

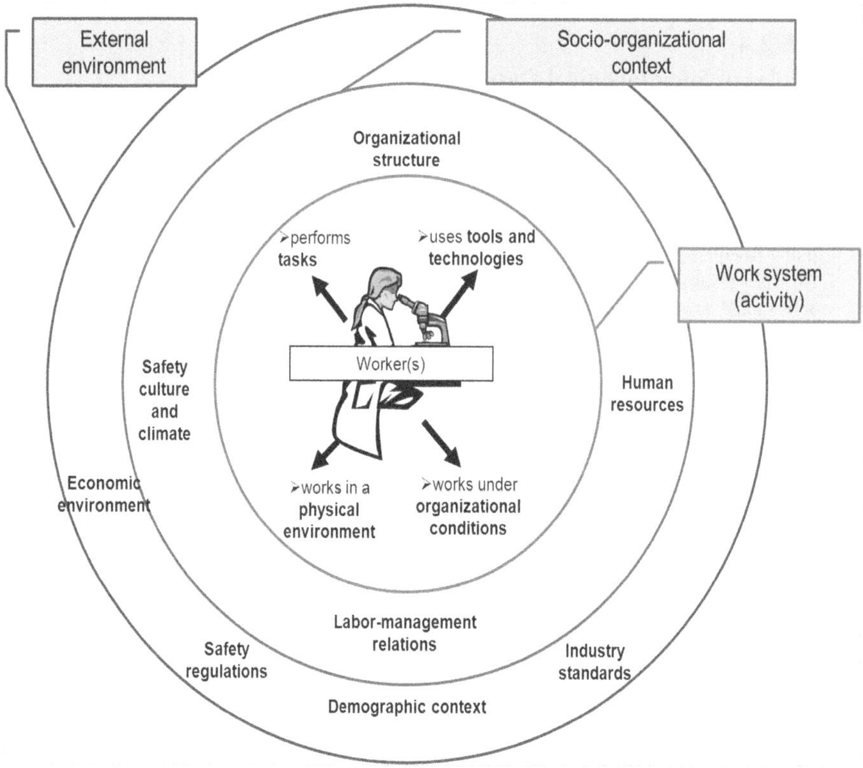

FIGURE 2.1 Model of sociotechnical system for workplace safety (Carayon, Hancock, et al., 2015).

(Carayon, 2009; Smith & Carayon-Sainfort, 1989). The work system is embedded in a socio-organizational context with several components, such as organizational structure, safety culture and climate and labor management relations. The outer circle is the external environment that includes safety regulations, the demographic context, the economic environment and industry standards.

Other HFE and safety models use layered 'vegetables' to describe system levels, such as the 'onion' model of Wilson and Sharples (2015) and Bogner's (2007) 'artichoke' model. Moray's (2000) model of sociotechnical systems includes the following levels: physical devices, physical ergonomics, individual behavior, team and group behavior, organizational and management behavior, legal and regulatory rules and societal and cultural pressures. He defines characteristics for each level; for instance, the individual behavior level is characterized by the individual's educational level, decision-making, perception, attention, thought, memory, anthropometrics and motor skills. Rasmussen's (2000) model of sociotechnical system defines six interacting levels: work, staff, management, company, regulators-associations and government. The levels should not be looked at in isolation; Rasmussen insisted that interactions among the levels of the sociotechnical system should be analyzed.

It is important to look at interactions between system levels, for instance, using a 'meso' approach (Rousseau & House, 1994). Karsh and colleagues (Karsh & Brown, 2010; Karsh et al., 2006) identified multiple system levels for understanding patient safety: individual, group or team, organization and external environment. They also developed an HFE model for patient safety that emphasizes interactions between the micro and macro levels (Karsh et al., 2014). This 'meso-ergonomics' approach has been used to analyze various safety issues, including safety violations during medication administration and HAI outbreak in hospitals (Karsh et al., 2014). This cross-level perspective is embedded in the description of sharp end and latent factors involved in safety (Cook et al., 1998; Reason, 1997). Work activities at the sharp end are influenced by the blunt end and the latent factors of the complex sociotechnical system.

2.4.5 Systems of Systems

Systems are connected to other systems and form systems of systems. As discussed earlier, this challenges the definition of boundaries, i.e. what is analyzed and considered in the systems analysis and problem-solving process aimed at improving health care safety. Jamshidi (2009) defines systems of systems as 'large-scale integrated systems that are heterogeneous and independently operable on their own, but are networked together for a common goal' (page 2). Large societal domains, such as health care, national security and transportation, represent systems of systems. Systems of systems pose unique difficulties for designing interventions and improving safety (Siemieniuch & Sinclair, 2014). Organizations within the different systems of systems are likely to have different structures, information systems, policies and resources; therefore, addressing problems at the interfaces between these organizations requires negotiations, conflict management, collaboration and coordination.

Health care is a system of systems (Clarkson et al., 2017). In their health and care journey across their life, people interact with multiple health care organizations, community organizations and care settings, including their home. Transitions between these systems can pose safety problems when information is not communicated, misunderstood or limited. Therefore, addressing health care safety requires attention to analyze and improve systems within specific organizations (e.g. a hospital or an ED) as well as across and between organizations. This concept of systems of systems is further developed in Chapter 3 on SEIPS with the SEIPS 3.0 model of patient journey (Carayon, Wooldridge, et al., 2020).

2.5 SYSTEMS THINKING AND HFE

A landmark paper by the International Ergonomics Association (IEA) Future of Ergonomics Committee defines three key characteristics of HFE (Dul et al., 2012):

- systems approach,
- human-centered design and
- improvement of performance and well-being

HFE is a systems discipline that aims to improve systems with people, such as work systems. According to Wilson (2014), 'Ergonomics/human factors is, above anything else, a systems discipline and profession, applying a systems philosophy and systems approaches' (page 5). I first describe the macroergonomics movement in the United States and the emergence of systems ergonomics in the United Kingdom; both provide the conceptual foundation for HFE systems approaches, including SEIPS. I then describe Wilson's (2014) features of HFE systems approaches and their applications to health care.

2.5.1 Macroergonomics and Systems Ergonomics

Around the world, the HFE discipline has evolved to embrace systems thinking and broaden the work systems to integrate organizational characteristics. This can be seen, in particular, in the writings of Klaus Zink (2000, 2006) in Germany and of Laerte Sznelwar (Bolis & Sznelwar, 2016; Sznelwar et al., 2008) in Brazil. In this section, I describe two specific efforts, one in the United States (macroergonomics) and a second one in the United Kingdom (systems ergonomics).

2.5.1.1 Macroergonomics in the United States

The Select Committee on the Future of Human Factors – 1980–2000 of the US Human Factors Society (later renamed Human Factors and Ergonomics Society or HFES) identified major societal changes (e.g. technology breakthroughs, world competition, demographic shifts), which challenged the future of HFE with its focus on 'micro-ergonomics' (Hendrick, 1991). In contrast, macroergonomics emerged from the need to integrate organizational design and management (ODAM) in HFE research and practice. At the micro-ergonomic level, HFE addresses interfaces between people and machines, physical environments, software or jobs, whereas macroergonomics focuses on the overall work system, including its organizational characteristics (Hendrick, 1991). Macroergonomics is based on the STS Theory and adopted and adapted concepts of sociotechnical systems, joint optimization of technical and social subsystems and human-centered design (Hendrick & Kleiner, 2001). Kleiner's (2006, 2008) model of sociotechnical system is composed of five interacting elements: technical subsystem (how work is performed), personnel subsystem (who performs the work), internal environment (psychosocial and physical), organization and management (how organization is designed) and external environment (sub-environments such as political and financial).

Macroergonomics also offers methods for better integrating HFE in system analysis, design and implementation, such as MacroErgonomic Analysis and Design (MEAD) (Hendrick & Kleiner, 2001; Kleiner, 2006) and participatory ergonomics (Noro & Imada, 1991). Macroergonomic methods have been implemented in health care, including MEAD for designing a medical event reporting system (Kleiner, 2007), participatory ergonomics for improving occupational health and safety of hospital staff (Bohr et al., 1997; Evanoff et al., 1999), variance analysis of diagnostic testing process (Hallock et al., 2006) and role network analysis for improving prevention of VTE among hospitalized patients (Hundt et al., 2017).

2.5.1.2 Systems Ergonomics in the United Kingdom

The original impetus for systems ergonomics came from the observation that HFE knowledge should be better integrated into systems design. Singleton (1967) used the term systems ergonomics in contrast to 'traditional ergonomics' to describe how HFE fits in systems design processes. How to ensure that HFE knowledge is used in the early stages of systems design has remained a concern and has been discussed by many HFE experts (Meister & Enderwick, 2001; Sinclair, 2007). According to Wilson (2014), 'systems ergonomics examines, accounts for and enhances the design of a system, and people's interactions with it, rather than concentrating on an individual part of it' (page 6). Wilson describes features of systems ergonomics that are discussed in Section 2.5.2.

A 2009 review of the use of systems ergonomics in patient safety research shows that many studies give lip service to systems thinking and HFE (Waterson, 2009). A recent review of a selected group of systems HFE methods shows an increase in the uptake of these methods in health care (Jackson et al., 2025); however, this review did not include many systems ergonomics methods, such as those discussed by Waterson et al. (2015). Further attention is necessary to understand interconnections between action, decisions and policies at multiple systems levels and how these are influenced by multiple networks of system characteristics and factors.

2.5.2 Wilson's Features of HFE Systems Approaches

Wilson (2014) defines six features of HFE systems approaches:

- systems focus,
- context,
- interactions,
- holism,
- emergence and
- embedding.

First, the focus of analysis should be on a system, its subsystems, the system in which it is nested, and the connected systems, i.e. systems of systems. These systems are sociotechnical systems as they include people interacting with technologies and many other elements. Second, human activities occur in a context, setting or work system; therefore, we need to analyze human activities 'in the wild' or real settings. Third, a system is composed of interacting elements; the interactions are more important than the elements themselves. Actually, in an earlier paper, Wilson (2000) suggests that the HFE discipline and profession are about designing 'interactions', such as temporal and spatial interactions, cooperation interactions or interactions with a technology. Fourth, holism is a concept that applies to the person (person as a whole), to the system (the whole system) and to the outcomes (e.g. physical, mental and emotional outcomes). This holistic approach fits with my concept of health care safety that combines safety for patients and safety for health care workers, and various dimensions of safety, well-being and performance. Fifth, Wilson (2014) describes

TABLE 2.5
Features of HFE Systems Approaches (Wilson, 2014) – Examples in Health Care Work Systems

Features of HFE Systems Approaches	Examples in Health Care Work Systems
Systems focus	Analyzing the work system of a specific job or role, e.g. a nurse in an intensive care unit, an emergency physician, health management of a patient with chronic disease. Analyzing a care process embedded in a distributed work system, e.g. hospital discharge process, diagnostic process.
Context (or work system)	Understanding the actual care setting and its multiple elements and dimensions.
Interactions	Interactions between members of the care team. Interactions between a patient and a technology (e.g. infusion pump, patient record).
Holism	Understanding physical, cognitive, social and emotional characteristics of people as they interact with other elements of the work system. Holistic perspective on safety that combines patient safety and health care worker safety.
Emergence	Expecting surprises when a new system or a redesigned system is in use. Understanding how people adapt to barriers in the work system (e.g. work-arounds); strategies can be hazardous but can also be sources of innovation and improvement. Looking for ways that people adapt the work system and innovate in their use of technologies and other elements.
Embedding	Facilitating and supporting the embedding of HFE professionals in health care organizations and other organizations involved in health and health care (e.g. health IT companies).

three implications of emergence for HFE: (a) HFE should be better integrated in design processes; however, we should always expect surprises once the new system is in use; (b) people develop strategies when faced with poor designs, barriers and constraints; and (c) people are creative and ingenious and can unexpectedly find new ways of using a technology or doing the work. The sixth feature of HFE systems approaches is about HFE in practice and embedding HFE professionals in organizations. Table 2.5 shows how the six features apply to health care work systems.

In Chapter 3 on SEIPS, the four features of systems focus, context (work system), interactions and emergence are further discussed. Holism is largely discussed in Chapter 1 when I propose the concept of health care safety or the combination of patient safety and health care worker safety. An important question remains about how to embed HFE within health care, i.e. the sixth feature of HFE systems approaches according to Wilson (2014).

2.5.2.1 Embedding HFE in Health Care

HFE can be considered as an innovation for improving health care safety (Carayon, 2010). Facilitators for the adoption of HFE in health care safety include the fact that HFE-based interventions have demonstrated a positive impact on performance, thereby facilitating the work of health care staff and improving the adoption of HFE, which can also be supported by local champions. On the other hand, multiple issues and challenges regarding integrating an innovation such as HFE exist, e.g. a lack of systems thinking, a lack of understanding of the benefits of HFE and the complexity of HFE.

Integrating HFE in health care is also about potential cultural clashes between HFE and its systems thinking, on the one hand, and health care culture, on the other hand (Carayon & Xie, 2011). We need to pay attention to professional cultures and expectations on both sides, i.e. HFE and health care; this can be done through education and training of health care professionals (Vosper et al., 2018) and leaders. HFE professionals who are embedded in health care settings represent another powerful mechanism for applying HFE concepts and methods to improve health care safety. Numerous challenges exist for effectively integrating HFE professionals in health care (Catchpole et al., 2021; Perry et al., 2021). On the other hand, successful examples of embedding HFE professionals in health care organizations exist, such as HFE in a large health care system in Portugal (Santos, 2020) and HFE in pediatric radiology in a US health care system (Larsen et al., 2024).

We need to support and enhance our commitment to HFE in health care safety (Bowie et al., 2025; Gurses, Ozok, & Pronovost, 2012). HFE systems approaches such as SEIPS are useful vehicles for achieving this objective. Increasingly, health care leaders and policymakers recognize that work systems (or sociotechnical systems) are complex; therefore, we need to propose approaches that focus on the entire system and consider interactions with other parts and levels of the system. In Chapter 3, I present SEIPS, an HFE-based systems approach that has been widely used in health care.

3 SEIPS or Systems Engineering Initiative for Patient Safety

This chapter is dedicated to describing SEIPS, its historical roots and evolution, its characteristics and features and the family of SEIPS models. Chapter 4 provides information on the 'how-to' of SEIPS; it describes different ways that SEIPS can be used in HCD to improve health care safety.

SEIPS stands for Systems Engineering Initiative for Patient Safety. It has been pronounced in different manners; at the University of Wisconsin-Madison, United States, we call it 'seeps'. The original SEIPS model published in 2006 (Carayon, Hundt, Karsh, et al., 2006) was developed by an interdisciplinary team at the University of Wisconsin-Madison, United States, that was awarded a grant from the Agency for Healthcare Research and Quality (AHRQ). This AHRQ grant for a Developmental Center for Evaluation and Research in Patient Safety (DCERPS) was administratively located in the College of Engineering and built on health systems engineering expertise in the Department of Industrial and Systems Engineering at the University of Wisconsin-Madison. This DCERPS became known as the SEIPS program as a group of researchers and practitioners from various disciplines (e.g., industrial and systems engineering, HFE, nursing, medicine, pharmacy, health services research, health informatics) collaborated to develop projects aimed at addressing complex health care safety problems. The SEIPS program developed over the years as we built collaboration across disciplines, universities, and health care organizations, and we created SEIPS courses and offered other learning opportunities.

Over the years, researchers affiliated with the SEIPS program have expanded and deepened their health care safety research. The expansion of SEIPS has helped to address multiple patient safety domains, such as medication safety (Carayon, Wetterneck, et al., 2021; Carayon, Wetterneck, Cartmill, et al., 2014), HAIs (Acher et al., 2015; Yanke et al., 2014), and diagnostic safety (Carayon, Hoonakker, et al., 2020). SEIPS research has also been expanded to go beyond the work of health care professionals and workers, and to understand the work of patients and care partners from a work system perspective (Holden et al., 2020; Holden, Schubert, & Mickelson, 2015; Valdez et al., 2024; Werner, Ponnala, et al., 2020). The SEIPS program has led to the creation of various SEIPS models, which I call the 'SEIPS family'.

The SEIPS family involves multiple authors and collaborators from various disciplines and horizons. Many are anchored in HFE and other engineering disciplines or domains (e.g. industrial and systems engineering, operations management). Other authors and collaborators come from clinical and health care domains; e.g. nursing, pharmacy, critical care, internal medicine, pediatrics, infection control and

 DOI: 10.1201/9781003480310-3

prevention, health informatics and health services research. The SEIPS family and its projects are characterized by a combination of research and practice, often using an action-research paradigm where an intervention is designed, implemented and evaluated. Several case studies described in Chapter 5 are examples of this cycle of design–implementation–evaluation–learning activities with both practical and research characteristics. SEIPS has been disseminated via publications, presentations, and projects, as well as training and education programs, such as the SEIPS short courses and various courses at the University of Wisconsin-Madison in the United States, e.g. a course on medication safety jointly offered by the School of Pharmacy and the College of Engineering (Karsh et al., 2005).

3.1 SEIPS FAMILY

SEIPS has evolved into a family of models:

- the original SEIPS model published in 2006 (Carayon, Hundt, Karsh, et al., 2006),
- an updated version of the 2006 model published in 2014 (Carayon, Wetterneck, Rivera-Rodriguez, et al., 2014),
- SEIPS 2.0 published in 2013 (Holden et al., 2013),
- SEIPS 3.0 published in 2020 (Carayon, Wooldridge, et al., 2020), and
- SEIPS 101 published in 2021 (Holden & Carayon, 2021).

Table 3.1 summarizes the core components of the SEIPS models and their unique features and characteristics. Figures 3.1–3.4 show the graphical representations of the SEIPS models, starting with the original 2006 version and the updated 2014 version, continuing with SEIPS 2.0 and SEIPS 3.0, and finally SEIPS 101.

The SEIPS models are anchored in the HFE discipline as the person is at the center of the work system, and the aim is to improve multiple objectives, in particular, patient safety and health care worker safety and well-being. The various versions of the SEIPS models help to answer the question of how to design work (e.g. work of physicians or nurses, teamwork, patient work) in order to achieve benefits for both patients and workers, as well as health care organizations and other entities in communities and society at large.

BOX – CORE COMPONENTS OF SEIPS

The core components of the SEIPS family are:

1. The work system model with multiple interacting elements and the person at the center of the system.
2. Multiple outcomes for patients and care partners, health care workers and health care organizations.
3. Feedback loops between work systems, processes and outcomes.

TABLE 3.1
Family of SEIPS Models

	Original SEIPS Model	Updated SEIPS Model	SEIPS 2.0 Model	SEIPS 3.0 Model	SEIPS 101 Model
References	Carayon, Hundt, Karsh, et al. (2006)	Carayon, Wetterneck, Rivera-Rodriguez, et al. (2014)	Holden et al. (2013)	Carayon, Wooldridge, et al. (2020)	Holden & Carayon (2021)
Core components	Work system model with interacting elements and person at the center Multiple outcomes for patients/care partners and workers/professionals Feedback loops				
Unique features and evolution	A combination of work system model of Smith & Carayon and the SPO model of Donabedian	Addition of the external environment	Description of patient work, the work of health care professionals and collaborative work. Configuration	Focus on the patient journey	Dissemination of SEIPS
Work system elements	Person Task Tools and technology Environment Organization	Person Task Tools and technology Environment Organization External environment	Person(s) Tasks Tools and technology Internal environment Organization External environment	Work system: • Care team • Task • Tools and technologies • Physical environment • Organizational conditions Socio-organizational context External environment	PEET: • **P**eople • **E**nvironments: physical, socio-organizational, external • **T**asks • **T**ools
Role of patients	Patient as a person at the center of the work system. Patient also involved in work system elements: e.g. nursing performing a task (e.g. administering medication) with a patient		Distinction between the different types of work system with different involvement of patient and health care professional	Focus on the patient and their journey. Explicit mention of caregivers and other members of the care team that surrounds the patient	Patient as people in the work system

Process	Process = how care is provided, delivered and managed: • Care process • Other processes, such as information flow, purchasing, maintenance, cleaning Process integrates all work system elements. Temporal nature of processes	Work processes: physical, cognitive, social/ behavioral Processes represent different types of work: • Professional work • Collaborative professional/ patient work • Patient work	Work system and process are intertwined and represented as the patient journey.	*Same as SEIPS 2.0*
Outcomes	Patient outcomes: quality of care and patient safety. Employee and organizational outcomes: job satisfaction and other attitudes, job stress and burnout, employee safety and health, turnover, organizational health (e.g. profitability). Linkages between patient outcomes and employee/organizational outcomes	Work outcomes for patients, professionals and health care organizations. Outcomes can be: • desirable/undesirable • distal/proximal	Patient safety and other outcomes (e.g. stress, health, performance) for: • patients • caregivers • clinicians • health care organizations	Work outcomes can be: • desirable/undesirable • distal/proximal. Balance of outcomes across various stakeholders (e.g. patients, informal caregivers, clinicians, health care organizations)
Feedback loops	Feedback loops from: • processes➔work system • outcomes➔work system Feedback loops represent pathways to design or redesign the work system. Feedback loops represent strategies of health care workers in reaction to the work system.	Feedback loops from: • processes➔work system • outcomes➔work system • outcomes➔processes Feedback loops represent adaptation: • anticipated/unanticipated • short- or long-lasting • intermittent or regular	Feedback loops from: • outcomes➔patient journey Feedback loops represent adaptation, improvement and learning.	Feedback loops from: • processes➔work system • outcomes➔work system • outcomes➔processes

Adapted and updated from Carayon et al. (2021).

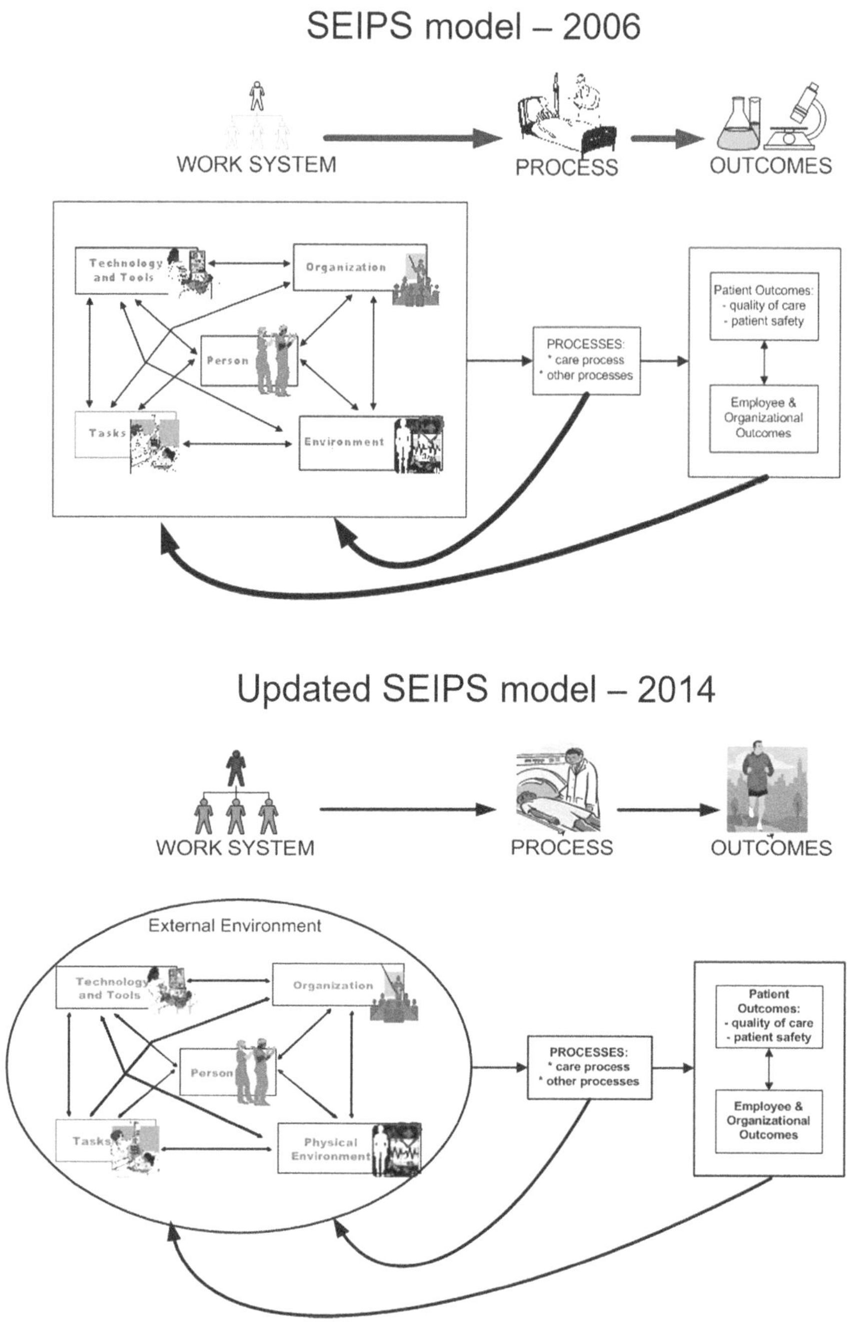

FIGURE 3.1 Original SEIPS model of work system and patient safety (Carayon, Hundt, Karsh, et al., 2006) and Updated Version (Carayon, Wetterneck, Rivera-Rodriguez, et al., 2014).

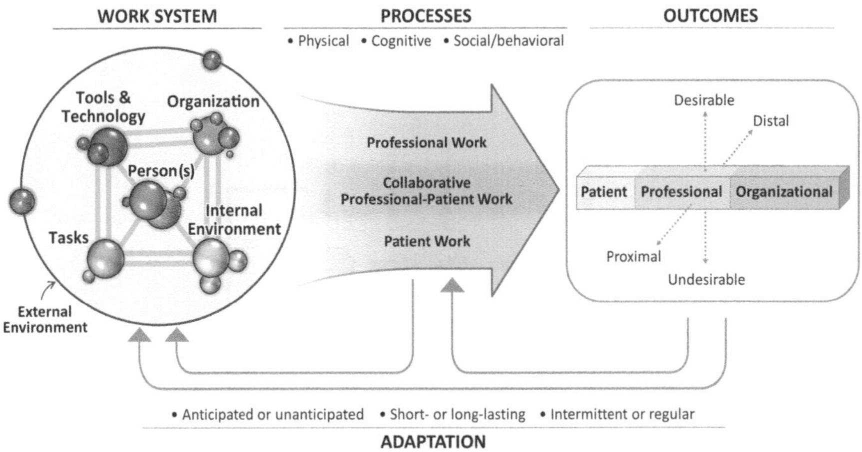

FIGURE 3.2 SEIPS 2.0 model (Holden et al., 2013).

The various members of the SEIPS family have unique features and specific emphases; see Table 3.1. For instance, the SEIPS 2.0 model distinguishes between the work of health care professionals, the work of patients and the collaborative work between health care professionals and patients (Holden et al., 2013). The SEIPS 3.0 model focuses on the patient and their journey while interacting with multiple health care organizations and settings (Carayon, Wooldridge, et al., 2020). It goes beyond the focus on episodic care and helps to address care transitions between settings as the patient 'travels' through their health and care journey. The question of which model to use is addressed later in Section 3.6 of this chapter.

3.2 WORK SYSTEM MODEL

As described in Section 3.1, the SEIPS models have core components, particularly the work system model. The work system model can be used to describe, analyze and improve work and outcomes in any organization or domain (not just health care). In line with the HFE discipline, the person is at the center of the work system, and the work system should be designed to support the person and not hinder their ability to do their work.

3.2.1 Historical Perspective of the Work System Model and the Balance Theory

In the late 1980s, (the late) Michael J. (Mike) Smith and I developed the work system model as a reaction to many models and theories, which at that time were narrowly focusing on specific aspects of work and ignoring the 'systemic' aspects of work. For example, in the field of job design, the Job Characteristics Theory focused on five job characteristics: variety, autonomy, task identity, significance and feedback (Hackman et al., 1975). In the field of job stress, the Job Strain model of Karasek and colleagues initially focused on two aspects of work: job demands (including workload) and

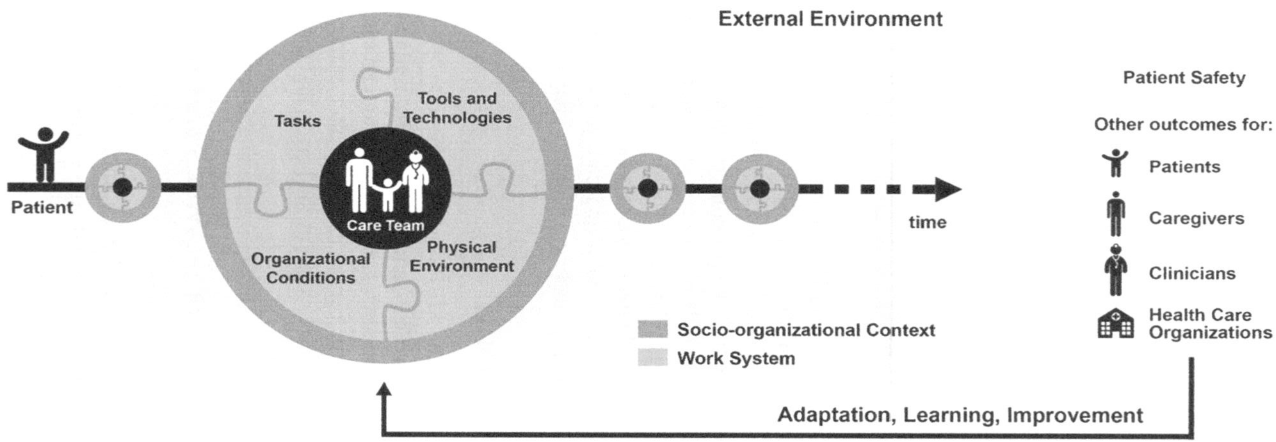

FIGURE 3.3 SEIPS 3.0 model of patient journey (Carayon, Wooldridge, et al., 2020).

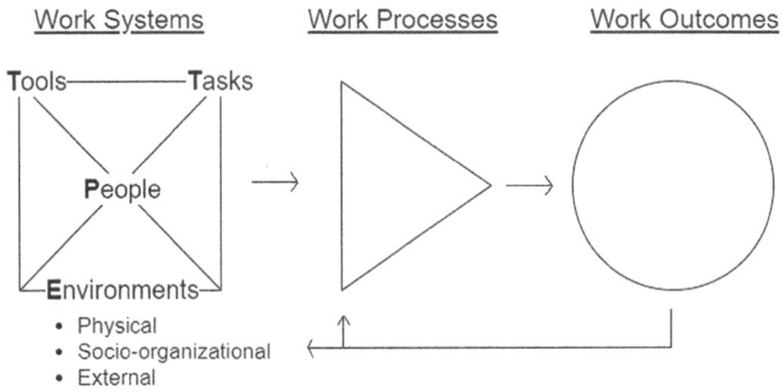

FIGURE 3.4 SEIPS 101 model (Holden & Carayon, 2021).

job decision latitude (including job control); the model was later expanded with the addition of social support (Karasek, 1979; Karasek & Theorell, 1990).

As explained in Section 2.5 of Chapter 2, in the 1990s, macroergonomics emerged as a topic of discussion within the Human Factors Society (later renamed Human Factors and Ergonomics Society or HFES) (Hendrick, 1991). Macroergonomics represented a major step toward integrating systems thinking in HFE, in particular concepts from the Sociotechnical Systems (STS) Theory (Hendrick & Kleiner, 2001). Until macroergonomics emerged at the end of the 1990s (Hendrick, 1991), HFE focused on the microergonomic aspects of work, such as physical design of workplaces, environments and tasks or design of equipment and devices, with little attention to the broader organizational and social context in which work is performed (Zink, 2000).

These various approaches to work design anchored in the fields of job/organizational design and job stress and the HFE discipline tended to be narrowly focused as they addressed specific aspects of work and a limited number of aspects of work. In addition, different bodies of literature were developed in silos and did not recognize overlapping concepts. For instance, autonomy was a key aspect of work in the job design literature and in the STS Theory, whereas in the job stress literature, the concept of job control was emphasized; autonomy and job control are similar concepts, but the connection between the two concepts was not recognized. In the 1980s, a few authors realized that various bodies of literature addressed the problem of job or work design and proposed models with a broader set of job characteristics. For instance, Campion and Thayer (1985, 1987) described four different bodies of literature that focused on different aspects of work design: (1) mechanistic approach (e.g. specialization, simplification, repetition), (2) motivational approach (e.g. autonomy, task variety, social interaction, feedback), (3) biological approach (e.g. lifting, sitting, noise), and (4) perceptual/motor approach (e.g. display design, information requirements, memory requirements). Parker and her colleagues (Parker, 2014; Parker et al., 2001) pursued this line of research and proposed expanded models of work design, including the SMART (Stimulating, Mastery, Agency, Relational, Tolerable demands) model (Centre for Transformative Work Design, 2025). The SMART model has been applied to the design of health care jobs, such as clinicians involved in the diagnosis of rare diseases (Hay, Klonek, et al., 2020).

In contrast with narrowly focused approaches to work design and in line with integrated approaches to work design and the emergence of macroergonomics, Mike Smith and I introduced the Balance Theory and the associated work system model in a 1989 paper published in the *International Journal of Industrial Ergonomics* (Smith & Carayon-Sainfort, 1989).

3.2.2 Work System Design and the Balance Theory

The main goal of our 1989 paper was to describe a systems approach to designing work, or what we call the Balance Theory (Carayon, 2009; Smith & Carayon-Sainfort, 1989). The Balance Theory involves two core principles for work system design:

1. eliminating the negative aspects of the work system and
2. balancing the work system through compensation and overall system balance.

Given various organizational or technological constraints, it may not always be possible to eliminate all or some of the negative aspects of the work system (first principle). Therefore, alternative strategies consist of either identifying aspects of the work system that can compensate for the negative aspects or ensuring overall system balance with significant positive aspects that counterbalance the impact of the negative aspects (second principle).

In our 1989 publication (Smith & Carayon-Sainfort, 1989), the work system model was presented as a group of interacting elements that determine working conditions, contribute to loads on the workers and provide resources for achieving individual and organizational goals. The elements of the work system can be described and characterized using concepts, models and theories of job/organizational design, job stress and HFE, such as those described above. Improving the design of work systems requires broad knowledge in work design that is found in the various connected fields of job/organizational design, job stress and HFE. This knowledge helps us to identify positive and negative aspects of the work system. For instance, the job/organizational design literature tells us that 'good' jobs are those with variety and opportunities for learning and problem-solving. We also know from the physical HFE literature that repetition should be avoided as it puts physiological demands on the body and can lead to injury. Therefore, a principle of work system design is to reduce or eliminate repetitive tasks and enhance task variety. The literature is replete with work design principles; some are described at the end of this chapter and others are discussed in the case studies presented in Chapter 5.

In work system design projects aimed at improving health care safety, we need to call on deep and broad knowledge of work design principles as well as information from data collection and analysis to understand what happens in the actual workplace. We can then identify work system barriers and facilitators through a combination of deduction (i.e. concepts, models and theories of work design) and induction (i.e. data collection and analysis of the actual work system).

We first introduced the concept of work system barriers and facilitators at a conference of the European Association of Work and Organizational Psychology, where we presented data from our outpatient surgery study and our study of ICU nursing workload (Carayon et al., 2005). We built on research by Peters and O'Connor (1980) on situational constraints and by Brown and Mitchell (1991) on performance obstacles. Barriers are factors in the work system that hinder performance and negatively affect quality of working life; on the other hand, facilitators are factors in the work system that support and facilitate performance and positively affect quality of working life. Our focus is on understanding what happens at the sharp end, and the experience of health care workers and patients regarding the care process, its constraints, its resources and its opportunities for improvement. Whereas some research in health care has focused on what does not work, e.g. operational failures (Tucker & Spear, 2006), we are interested in identifying both minuses and pluses in the work system, i.e. work system barriers and facilitators.

3.2.3 Work System Elements

The original publication on the work system model and the associated Balance Theory described the work system as a set of five interacting elements: *the person (at the center), tasks, technology, environment and the organization* (Carayon, 2009; Smith & Carayon-Sainfort, 1989). As shown in Table 3.1, the work system elements vary across the SEIPS family. The original five elements described in our 1989 publication are also part of the SEIPS model of 2006 (Carayon, Hundt, Karsh, et al., 2006). In the updated SEIPS model of 2014 (Carayon, Wetterneck, Rivera-Rodriguez, et al., 2014), we added the external environment to highlight the open system feature of the work system. The work system interacts with the external environment: it is affected by and affects the external environment. In the SEIPS 2.0 model (Holden et al., 2013), the physical environment is called 'internal environment' and includes characteristics of lighting, noise, workspace and layout. See Table 3.2 for a description and examples of work system elements.

The SEIPS 3.0 model (Carayon, Wooldridge, et al., 2020) uses the same work system elements included in the original SEIPS model (Carayon, Hundt, Karsh, et al., 2006) and the updated version of the SEIPS model (Carayon, Wetterneck, Rivera-Rodriguez, et al., 2014): *people, tasks, tools and technology, physical environment, organizational conditions and external environment.* It distinguishes between two system levels: the local work system (i.e. what happens at the 'sharp end') and the socio-organizational context, e.g. a health care organization or a home setting. This distinction of the sharp end and the socio-organizational context is aligned with the concept of system levels; see the discussion of system levels in Chapter 2. The SEIPS 3.0 model integrates the multi-level model of the sociotechnical system for workplace safety (Carayon, Hancock, et al., 2015), which shows how the work system is embedded in a socio-organizational context, which in turn is influenced by and influences the external environment (see Figure 2.1 in Chapter 2). This multi-level model fits with the concept of 'sharp end' and 'latent factors' in safety sciences (Cook et al., 1998; Reason, 1997) as described in Section 2.4.4 on system levels in Chapter 2 as well as other multi-level models of HFE and safety; e.g. the onion model of Wilson and Sharples (2015), the artichoke model of Bogner (2007), and the sociotechnical model of Moray (2000).

TABLE 3.2
Work System Elements in SEIPS

Work System Elements	Description and Examples
People	Health care staff: physician, nurse, pharmacist, therapist, technician, clerk, …. Patient Care partner Physical, psychological and psychosocial characteristics of people
Tasks	List of tasks and their characteristics: e.g. frequency, duration, repetitiveness Workload, work pressure Job control, autonomy, flexibility Job content, meaning of work, feedback Ambiguity, uncertainty
Tools and technology	List of tools and technology Usability Accessibility Physical characteristics of tools and technology: e.g. dimensions, strength necessary
Physical environment	Layout and workspace Ambient environment: e.g. lighting, noise, temperature, humidity
Organization	Team work Communication and coordination Social support Work schedules Training Organizational resources Management style and leadership Safety climate and culture Organizational climate and culture
External environment	Socioeconomic environment Workforce Regulations and policies Professional societies Patient organizations Local community

The SEIPS 101 model (Holden & Carayon, 2021) is a simplified version aimed at improving the dissemination of SEIPS and facilitating its use. In SEIPS 101, the work system is comprised of: *people, environments, tasks and tools* (or PETT). The environments are broadly conceptualized and include physical aspects (e.g. noise, lighting, layout of workspace), socio-organizational aspects (e.g. organizational culture, organization of the home environment) and characteristics of the external environment (e.g. regulations, legal environment).

The work system model is a generic model that describes any type of work. The elements of the work system vary slightly across the various members of the SEIPS family. Describing each work system element is important, but more importantly, we need to consider the system interactions and how these system interactions in the care process influence health care safety.

3.3 WORK SYSTEM AND PROCESS

The work system model is a core component of the SEIPS family, but SEIPS is more than the work system model. SEIPS integrates the work system model with the care process, outcomes for patients, care partners, health care workers and health care organizations and feedback loops (i.e. the core components of SEIPS; see Table 3.1 and Box).

In HFE, the focus is often on the task as demonstrated by multiple methods of task analysis (Kirwan & Ainsworth, 1992). The term process is sometimes used to describe an element of a task. For instance, a task involves physical, cognitive and psychosocial processes. This is not how I use the term process. To me, a process is bigger than a task; a process is actually a group of tasks. This use of the term 'process' is aligned with the industrial and systems engineering discipline, i.e. my own background. In health care safety, we need to move beyond our focus on task and expand our analysis 'from tasks to processes' (Walker & Carayon, 2009).

3.3.1 Process in SEIPS

In the original 2006 SEIPS publication (Carayon, Hundt, Karsh, et al., 2006), we described how the SEIPS model was created as a combination of the Structure–Process–Outcome (SPO) model of Donabedian (1966, 1978) and the work system model of Smith and Carayon (Carayon, 2009; Smith & Carayon-Sainfort, 1989). Avedis Donabedian was a thought leader in health services research who developed the SPO approach for the assessment of health care quality. He defined three types of care quality measures: structural measures (e.g. adequacy of health care facilities, qualifications of staff), process measures (e.g. justification of diagnosis and therapy, evidence of preventive management, coordination and continuity of care) and outcome measures (e.g. illness, mortality, longevity) (Donabedian, 1966). In creating the SEIPS model, we adapted the SPO model in three different ways:

1. We replaced the 'S' (structure) component with the work system model.
2. The 'P' (process) consists of the steps of patient care, i.e. what is done to and with the patient.
3. In line with the HFE discipline, we expanded the 'O' (outcomes), and included multiple outcomes related to the safety, health, performance and well-being of both patients and health care workers.

In creating the SEIPS model, we included processes connected to the care process (e.g. supply chain management) and added feedback loops in order to represent mechanisms of adaptation, improvement and learning.

3.3.2 Work System and Care Process are Intertwined

In the various graphical representations of the SEIPS models (see Figures 3.1–3.4), the work system and the care process are represented separately as the work system influences the process. It is important to understand that the work system and the process are intertwined: we cannot separate them as they are intrinsically related.

A care process can actually be described using the work system and its elements. In a 2004 publication, we described the case of a wrong-site surgery as a flow of *tasks* (e.g. preparing the patient for surgery, postop education) performed by various *individuals* (e.g. surgeon) using various *tools and technologies* (e.g. medical record); the tasks are performed in various *physical environments* (e.g. anesthesia preop area, operating room) and under *organizational conditions* (e.g. surgery team, coordination between anesthesia, surgery and primary care) (Carayon et al., 2004). Characteristics of the work system elements were embedded in the process and led to a wrong-site surgery; e.g. noise that affected communication between the surgeon and the hearing-impaired patient led to mis-identification and a wrong-site surgery.

The work system model is typically visualized as a group of interacting elements. The care process can be considered as a temporal instantiation of the work system elements: the work system elements are combined, and they evolved over time in the care process. The SEIPS-based process modeling method developed by Wooldridge et al. (2017) integrates the work system elements in mapping the process. The resulting process map shows the *tasks* performed as they unfold chronologically. The process map is shown in a tabular format where rows represent time and columns represent *who* completes the *tasks*, where the tasks are done (*physical environment*), and/or the organizational unit where the tasks are performed (*organization*). The *tasks* are accompanied by icons that describe the *tools and technologies* used to perform the tasks. A SEIPS analysis often includes a description of barriers and facilitators. In the SEIPS-based process modeling method, the barriers and facilitators appear in the process map along the tasks; they are represented by red (barriers) and green (facilitators) dots. Examples of process maps constructed with the SEIPS-based process modeling method can be found in Wooldridge et al. (2017) in a study of three processes (i.e. pre-visit planning, patient outreach for panel management and checkout) in two primary care clinics and in Odberg et al. (2020) in a study of the medication administration process in a Norwegian nursing home. The SEIPS-based process modeling method clearly shows that the work system and the process are intertwined.

A care process often represents interactions among multiple work systems. Understanding the information flow, communication and coordination among these different work systems can provide important perspectives on how to improve health care safety. For instance, an in-depth analysis of the sterile processing of surgical instruments highlighted the interdependencies between various actors and units, including the Sterile Processing Department (SPD), the Operating Room (OR) and the delivery courier network (Segarra et al., 2024). A SEIPS 101 journey map showed the various steps of a surgical instrument tray through the work systems of the SPD, the OR and the courier network. Failures or problems at one step can lead to failures or problems at a subsequent step, such as missed or incorrect tray bar code scans that can affect coordination and delay delivery of surgical trays to the OR, potentially delaying a surgery. Cascades of disturbances or errors in work systems can affect safety in the process, especially if they are not detected, corrected or addressed. For other examples of temporal aspects of safety, see studies on medication error propagation described in Section 1.1.3 in Chapter 1. A process analysis often requires the analysis of multiple inter-connected work systems; this means analyzing each work system as well as their inter-relationships.

3.3.3 SEIPS Versus Quality Improvement

Quality improvement (QI) professionals use various methods to describe and analyze processes; many of them look similar to methods used in SEIPS projects. However, as emphasized by Bowie and Vosper (2024), SEIPS is anchored in HFE and has different assumptions than QI. According to HFE, people are at the center of work systems and processes should be designed to support their work and positive outcomes. Work systems are dynamic and interact over time and space as shown by the SEIPS 3.0 model (Carayon, Wooldridge, et al., 2020). Therefore, when describing the process of a patient's journey, we need to understand that multiple work systems are involved and interact with each other. For instance, various barriers and facilitators occur in the diagnostic process as the patient interacts with multiple health care work systems; e.g. making an appointment, getting diagnostic tests, communication between patient and clinician regarding the diagnosis and implementation of the treatment plan (Rasooly et al., 2024).

When introduced to the SEIPS model, QI professionals often compare the work system model to the Fishbone diagram. QI professionals typically use the Fishbone diagram to identify factors that contribute to the quality of care processes and outcomes. The Fishbone diagram, also known as the Ishikawa diagram, provides a visual representation of the contributing factors in various categories, such as people, process, equipment, materials, environment and management. For example, Graber et al. (2024) use the Fishbone diagram to describe and visualize four factors contributing to diagnostic errors: context of care (e.g. health IT, communication, lab & imaging resources), the clinician's reasoning process (e.g. knowledge deficit, incorrect synthesis), the patient and family perspective (e.g. inability to give complete history, misunderstandings, compliance problems) and the specific characteristics of the case (e.g. newly described or rare disease, urgency). As explained by Bowie and Vosper (2024), SEIPS is significantly different from the Fishbone diagram. SEIPS clearly provides a systems approach to health care safety. There are complex interactions between system elements that influence the care process and impact various outcomes; this is not part of the Fishbone diagram that describes factors separately.

Bowie and Vosper (2024) compare and contrast the philosophy and assumptions of SEIPS and the Fishbone diagram. For instance, SEIPS emphasizes system interactions that are dynamic and nonlinear, whereas the Fishbone diagram assumes linear cause and effect. Because SEIPS is anchored in the HFE discipline (Dul et al., 2012), SEIPS-based analyses of care processes aim at improving safety for both patients and health care safety.

3.4 INTERACTIONS IN DYNAMIC WORK SYSTEMS

In line with systems thinking, the work system model is composed of elements that interact with each other. As discussed in Chapter 2, system interaction is a key feature of a systems approach. In the context of the work system model and the Balance Theory (Carayon, 2009; Smith & Carayon-Sainfort, 1989), elements interact as they are combined, influence each other, are impacted by a change or intervention or change over time. The work system is dynamic as people change, learn, evolve, react and adapt; therefore, system interactions also change over time.

Graphical representations of the work system model show system elements that are connected to each other with double-headed arrows (see Figures 3.1–3.4). Of course, these graphical representations cannot do justice to the multiple ways that system elements interact. Interactions include not just interactions between two system elements; the interactions can include more than two elements and could go one way or both ways and are dynamic. A systems approach requires that one thinks in systems and can identify these complex system interactions. Thinking in systems can be supported by the use of the STEW (Systems Thinking for Everyday Work) principles (Fell et al., 2024; McNab et al., 2020); see Table 2.2 in Chapter 2.

Interactions, interconnections, interdependencies,…; these are all different terms used in the systems thinking literature. For instance, Meadows (2008) talks about interconnections as the relationships that hold the elements together. Others talked about interdependencies when applying SEIPS (Bowie & Vosper, 2024; Segarra et al., 2024). Wilson (2000) describes HFE as the discipline that allows us to understand and design complex interacting systems that include people. Interactions are a core feature of systems thinking and HFE systems approaches such as SEIPS. I distinguish between four types of system interactions in SEIPS; see Table 3.3 and subsequent Sections 3.4.1–3.4.4.

3.4.1 Interactions Between Work System Elements

The SEIPS model identifies specific elements of the work system; see Table 3.2. As discussed above, the different SEIPS models have slightly different versions of the elements. The elements are important as they draw attention to the entire set of characteristics, factors, constraints and resources involved in work. However, it is important that a SEIPS-based analysis goes beyond describing the elements and understands the interactions between system elements. The elements are not independent from each other. When analyzing adverse events in health care safety, we often find multiple work system interactions. For instance, Baartmans et al. (2023) analyzed 69 adverse events reported by 28 Dutch hospitals and found numerous interactions between work system elements, such as a sub-optimal patient room layout (*physical environment*) increasing workload and leading to adverse events. When, for instance, describing the activity of medication administration by a nurse, we need to understand the *tasks* performed by the nurse (e.g. placing an IV line), the *tools and technologies* used by the nurse to perform the tasks (e.g. infusion pump, syringe) and the *environment* where the tasks are performed (e.g. patient room in a hospital with multiple interruptions). The tasks are influenced by *organizational conditions* (e.g. availability of nursing colleagues to help with double-checking high-risk medications). This short example clearly shows multiple system interactions. The goal of SEIPS is to define these system interactions and understand how to improve the overall system performance and safety and well-being outcomes.

TABLE 3.3
Work System Interactions in SEIPS

Types of Work System Interaction	Description
Interactions between work system elements	• Work system elements interact. • These interactions produce barriers or facilitators that influence health care safety and can lead to adverse events. • Barriers and facilitators interact with each other.
Change in the work system	• A change may occur in a work system element or part of the system. • The change could be planned (e.g. an intervention aimed at improving the design of the physical environment) or unplanned (e.g. adaptation resulting from a crisis in the external environment, such as the COVID-19 crisis). • Changes are never contained to a single element or aspect of the work system. • Changes have systemic impact on the entire work system and its elements through multiple interactions.
Interactions between system levels	• System levels interact with each other, in particular the local work system and the socio-organizational context (see the SEIPS 3.0 model). • Top-down system-level interactions: what happens at the local level is influenced by decisions made at the socio-organizational level and in the external environment. • Bottom-up system level interactions: information flows from the local level (e.g. about something that works or something that does not work) and influences action at the socio-organizational level and in the external environment.
Dynamic nature of the work system	• As time passes by, elements (and their interactions) evolve and change. • Temporal changes in specific elements or interactions influence the entire work system through multiple interactions.

3.4.1.1 Configuration

All work system elements interact with each other; the possible interactions are, therefore, infinite. But, in a specific care process, it is possible that only a small network of work system elements is relevant for a specific health care safety problem. This is the objective of 'configuration', a key concept of the SEIPS 2.0 model (Holden et al., 2013). A SEIPS-based configuration analysis is shown in the form of configural diagrams where relevant work system elements and their specific elements are displayed. Configural diagrams have been used to compare work systems before and after an intervention aimed at improving the diagnosis of rare diseases (Hay, Klonek, & Parker, 2020), to identify work system barriers and show the misalignment of work system factors in two primary care nursing processes (Tarver et al., 2024) and to compare the ED work systems that influence disposition decision-making under low- and high-demand conditions (Rutkowski et al., 2025). Configuration is further discussed in Section 4.8.3 of Chapter 4 on Human-Centered Design and SEIPS in Action.

3.4.1.2 Balanced Work System

The work system elements as well as their interactions determine work system barriers and facilitators. In a health care safety improvement project, the goal should be to eliminate barriers and keep facilitators or create new facilitators. Per the Balance Theory (Carayon, 2009; Smith & Carayon-Sainfort, 1989), it may not always be possible to eliminate barriers because of constraints. In this case, the Balance Theory proposes compensatory balance, where a facilitator compensates for or mitigates the effect of a barrier, and overall system balance, where the facilitators have a greater impact than barriers. For instance, having some autonomy over when and how to perform tasks can help in managing workload; this is an example of compensatory balance.

In an empirical study, Wooldridge et al. (2024) identified two additional types of interaction between work system barriers and facilitators, besides compensatory and overall system balance: amplification and exacerbation.

Amplification: interactions between facilitators can enhance their individual positive effects and produce a greater combined positive effect. This interaction between facilitators can also occur between barriers. For instance, an organizational facilitator of effective and efficient training with a new technology can enhance the technological facilitator of a highly usable technology and produce an enhanced positive impact on the technology user's performance and satisfaction.

Exacerbation: a facilitator in the work system may actually exacerbate a barrier in the work system. This interaction between a facilitator and a barrier is contrary to the compensatory balance mechanism. The facilitator does not help in mitigating the negative impact of the barrier; on the contrary, the facilitator increases its negative impact. For example, a facilitator of 'convenience' may arise when a new health information technology is usable and portable and allows a physician to write prescriptions in multiple locations (e.g. at the bedside, in their office); this represents a facilitator that results from interactions between *the technology, the task and the physical environment* elements. This facilitator may actually enhance the barrier of negative work–life balance as physicians can write prescriptions at any time, including at home.

Interactions between work system elements, as well as the different types of interaction between work system barriers and facilitators (balance, amplification, exacerbation), should be considered in any work system redesign.

3.4.2 Change in the Work System

Changes in health care occur frequently; sometimes driven by the external environment (e.g. change in rules and regulations) or by organizational strategies (e.g. implementation of EHR technology to improve flow of information and clinical decision-making). These changes include organizational changes as well as multiple technological changes, such as the implementation of health information technologies (e.g. EHR, bar coding medication administration technology) and medical devices (e.g. smart infusion pump, surgical robots). It is important to recognize that any change will have an impact on the entire work system.

Over the years, I have often heard comments such as: 'Don't worry, this is just [*fill in the blank*]…' The assumption of a change being small and having little impact is unreasonable in light of systems thinking. Any change will have an impact on the entire work system. Therefore, whenever implementing a change, health care organizations should pay attention to its systemic impact; this should be done at multiple stages: when the change is designed, when it is implemented, when it is evaluated and when it is further improved. Too often, negative consequences of a change are called 'unintended consequences' as decision-makers argue that they did not intend for the change to produce challenges, negative outcomes or even harm. Let me digress and review and clarify the concept of unintended consequences.

BOX – UNINTENDED OR UNANTICIPATED CONSEQUENCES?

The concept of unintended consequences is largely used to describe the consequences of a change or an intervention that are not anticipated. Unintended consequences are often described as negative surprises related to the implementation of a change. For instance, Ash and her colleagues (2009) described the 'unintended consequences' of various types of health information technology (e.g., CPOE or Computerized Provider Order Entry), such as new work or more work, workflow issues, and new kinds of error.

Since the landmark paper of the sociologist Robert Merton on unanticipated consequences of purposive action (Merton, 1936), the term 'unintended consequences' has been used as a synonym of unanticipated consequences, and has largely replaced it (De Zwart, 2015). Unintended consequences are, then, described as consequences of the change or the intervention that occur after the implementation, but were not anticipated while planning the change or the intervention. De Zwart (2015) describes the problems associated with the conflation of unintended and unanticipated consequences. Unanticipated consequences are, of course, unintended, but unintended consequences may be either anticipated or unanticipated. Unfortunately, the conflation between unintended and unanticipated consequences does not do justice to the category of unintended consequences that are anticipated.

When planning a change, decision-makers anticipate consequences or outcomes from the intervention in the work system; for instance, when implementing CPOE, health care organizations aim to improve medication safety, in particular at the stage of medication prescription. Among the consequences, some are intended, i.e. the change is supposed to, e.g., improve medication safety; other consequences may be anticipated, but not intended (e.g. resistance to the change) and finally, some consequences are not anticipated nor intended (e.g., additional workload due to the change in workflow). Conflating unintended and unanticipated consequences may limit our ability for a critical approach to and analysis of the change (Parvin & Pollock, 2020).

If unintended consequences are all unanticipated, then we cannot discuss the possibility that some of these consequences could have been, would have been or should have been anticipated. According to Parvin and Pollock (2020), this may allow 'enthusiasts (of a particular policy, technology, or what have you) to abdicate responsibility for the perfectly foreseeable consequences of particular decisions, by positioning those consequences as outside their purview' (page 323). For instance, the 'unintended consequences' of health IT described by Ash and her colleagues (2009) look very similar to consequences that occur after the introduction of many other information technologies and that are well documented such as ironies of automation (Bainbridge, 1983) or automation surprises (Sarter et al., 1997). Maybe we should have known or should have anticipated these so-called 'unintended consequences of health IT'. In any case, now that we know about the so-called 'unintended consequences of health IT', how can health care organizations not consider them in their health IT implementation?

In a way, the conflation of unintended and unanticipated consequences may limit the imagination and creativity of technology (or any change) implementers as they abdicate their responsibility and call for their ignorance when the change has a negative unintended consequence. Therefore, in line with De Zwart (2015) and Parvin and Pollock (2020), I call for caution and clarification when the concept of 'unintended consequences' is used.

Going back to our discussion of change in the work system, it is clear that any change will affect the entire work system, its elements and their interactions. Some of these changes may be intended, and therefore anticipated; others may be unintended and unanticipated; and finally, others may be unintended but also anticipated. Whenever implementing a change, we should expect surprises, that is, changes in the work system and outcomes that we did not anticipate. Some changes may not be anticipated, but we should anticipate being surprised; this is related to the concept of emergence. Expecting surprises is in line with systems thinking. Therefore, whenever a change is implemented, we need to monitor and assess its impact; this can be done using the SEIPS model, e.g. identifying work system barriers and facilitators and assessing changes in the process and impact on the outcomes. The feedback loops are important in the SEIPS family; they are about adaptation, improvement and learning over time.

3.4.3 Interactions Between System Levels

A third type of interaction in the SEIPS family is about top-down and bottom-up interactions between system levels. As described in Section 2.4.4 of Chapter 2, understanding interactions between system levels is an important feature of systems thinking. System levels are clearly shown in the SEIPS 3.0 model as concentric circles (Carayon, Wooldridge, et al., 2020): the inner circle is the local work system and is composed of five elements (*the care team, tasks, tools and technologies, physical environment and organizational conditions*); the outer circle is the *socio-organizational context*. The two concentric circles interact with each other and also interact with the *external environment*.

What happens at the local level is influenced by the *socio-organizational context*, such as organizational decisions regarding work scheduling and training, and by the *external environment* (e.g. rules and regulations regarding work schedules for physicians-in-training). In return, the *socio-organizational context* may be influenced by the local work system. For instance, clinical staff at the 'sharp end' may provide suggestions for improving patient safety during executive walk-rounds (Frankel et al., 2003). These formal mechanisms organize interactions between the local level and the organizational level. For instance, in a study of the implementation and impact of executive walk-rounds in a hospital (Thomas et al., 2005), sample questions asked by executives included:

- Have there been any 'near misses' that almost caused patient harm but didn't?
- What aspects of the environment are likely to lead to the next patient harm?
- Is there anything we could do to prevent the next adverse event?
- Can you think of a way in which the system or your environment fails you on a consistent basis?
- What specific intervention from leadership would make the work you do safer for patients?

Several of these questions could be enriched by using SEIPS. For instance, the second question on 'aspects of the environment' that can lead to patient harm could be organized around the work system elements; this could provide further details on what works and what does not work at the local level, and what influences patient safety as well as worker safety. This information could then be used at the socio-organizational level by health care leaders and workers to make decisions regarding patient safety improvement strategies.

3.4.4 Dynamic Nature of the Work System

The work system is a dynamic system; it changes and adapts over time. The person at the center of the work system changes with experience, training, interactions with others, etc. The person also reacts and adapts to other elements of the work system. These changes and interactions occur over time and reflect the dynamic nature of the work system; emergent states of the work system result from these dynamic interactions.

Representation of time and the dynamic nature of the work system are both implicit and explicit in SEIPS. The dynamic nature of the work system is implicit as all elements may change over time, but these temporal changes are not clearly visualized in many graphical representations of SEIPS and the work system except for arrows between work system elements and the feedback loops. The feedback loops in the various SEIPS models explicitly represent time as information about outcomes is fed back to the work system and the process; this information can then be used for learning about what works and what does not work, and for identifying opportunities for improving the work system. These feedback loops also represent the mechanism of adaptation.

3.4.4.1 Adaptation

Adaptation is one of the key concepts in SEIPS 2.0 (Holden et al., 2013). Some adaptation is planned; for instance, the transition to telehealth is a planned adaptation to provide remote care and support access to care for patients in remote or isolated areas. Adaptations are also reactive; they represent strategies as people respond and adapt to aspects of the work system (Durso et al., 2014), such as work pressures (Page et al., 2024; Page et al., 2025). For instance, Tucker and colleagues (Tucker & Edmondson, 2003; Tucker et al., 2002) describe first-order problem-solving strategies of nurses who frequently experience operational problems or barriers, such as lacking supplies or medications.

In our research, we have described the various adaptation strategies used by health care workers when facing barriers, such as technology barriers experienced by care coordinators in charge of managing chronically ill patients (Carayon et al., 2019) and work system barriers experienced by tele-ICU nurses (Hoonakker & Carayon, 2018). In those studies, we define strategies as activities aimed at dealing with work system barriers. In the study of technology barriers experienced by care coordinators (Carayon et al., 2019), we identified three categories of strategies: (1) doing nothing or delaying performance, (2) working around the barrier and (3) tackling the barrier directly. In the study of tele-ICU nurses (Hoonakker & Carayon, 2018), we examined the entire range of work system barriers experienced by tele-ICU nurses and identified a wide variety of strategies, including work-around, taking little or no action and anticipating the problem or barrier. These strategies reflect various types of adaptation to and of the work system; they are dynamic interactions that unfold as the person works and reacts to the work system.

When reacting to a specific aspect of the work system, health care workers may change this aspect or other characteristics of the work system. Both work system barriers and adaptation strategies can, therefore, be mapped onto the work system. Watt et al. (2019) analyzed the safety of blood transfusion and the work and adaptations of clinical and laboratory workers involved in the blood transfusion process. They identified a range of 'triggers' (or barriers) in all work system elements and system adaptations made in response to the triggers. The system adaptations addressed all work system elements; in only 30% of the instances, the adaptation was made to the same element as the trigger. For instance, the physical separation between the room where the blood collection is done and the room for storing the blood created a barrier for the laboratory staff to monitor and help a nurse collecting blood. A simple two-way baby monitor was installed so that laboratory staff could hear what is happening in the collection room and intervene if a nurse needed help. This technology-based adaptation for supporting communication between the laboratory and clinical staff was in response to the issue of physical separation between the two work-rooms. This example shows a system adaptation in the *technology* element in response to a barrier in the *physical environment* element of the work system.

This study uses an interesting combination of SEIPS and the CARE (Concepts for Applying Resilience Engineering) model (Anderson et al., 2016). Adaptation strategies can provide ideas for work system redesign. Analyzing adaptation to the work system can be integral to learning and continuously improving health care safety.

3.4.4.2 Patient Journey Unfolds Over Time and Space

The SEIPS 3.0 model explicitly depicts time as the patient interacts with various work systems during their journey that unfolds temporally and spatially (Carayon, Wooldridge, et al., 2020). The SEIPS 3.0 model can help to understand how health care safety improves or deteriorates over time through an analysis of the various work systems involved in the patient journey or what Eason (2014) calls 'trans-organizational work systems'. For instance, a study by Bourne et al. (2023) analyzes medication safety in the journey of intensive care patients who transition to a hospital unit. An interview guide based on the SEIPS 3.0 model (Carayon, Wooldridge, et al., 2020) and the London Protocol (Taylor-Adams & Vincent, 2004) was used to collect data from 22 clinicians (e.g. physician, pharmacist) from the ICU and other units in multiple UK hospitals. Interactions between the ICU and hospital unit work systems influenced medication safety for ICU patients during their transition to a hospital unit. For instance, inter-team (ICU to hospital unit) communication was identified as a challenge to medication safety; this *organizational* factor was also affected by many other work system factors, such as the presence of e-prescribing systems available across the ICU and the hospital unit that supported the medication information flow. The SEIPS 3.0 model can be particularly useful to understand the dynamic nature of health care safety along the patient journey and the interfaces and care transitions between multiple work systems (Grimes, 2021).

3.4.5 Visualizing System Interactions

Visualization allows for the understanding of interactions such as those occurring in a complex work system and care process (Antonacci et al., 2021). A typical visualization of the work system model is a display of elements that are connected with arrows; the arrows are about interactions; see Figures 3.1–3.4. Drawing arrows between work system elements, between work systems in a care process, between multiple outcomes and/or between different components of SEIPS models is critical for emphasizing interactions. Multiple SEIPS-based methods help to visualize system interactions, such as:

- process map (Carayon et al., 2004)
- SEIPS-based process modeling (Wooldridge et al., 2017)
- SEIPS-based patient journey mapping (Wust et al., 2022)
- configural diagram (Holden et al., 2013; Tarver et al., 2024)
- role network analysis (Hundt et al., 2017)

Several of these methods are described in other chapters. Here, I describe one method aimed at visualizing the work of a care team: the role network analysis.

The role network analysis was specifically developed to analyze the team of *individuals* involved in a care process, their *tasks* and associated *tools and technologies* and the *individual and team work* (Hundt et al., 2017). Role network analysis belongs to the toolbox of the STS Theory (Pasmore, 1988) and is incorporated in the macroergonomic method of MEAD (Hendrick & Kleiner, 2001; Kleiner, 2006). We applied the role network analysis to understand the team involved in VTE prophylaxis management for hospitalized patients at different stages of the hospitalization (e.g. upon admission, during hospital stay) (Hundt et al., 2017). Ensuring appropriate VTE prophylaxis is key to avoid adverse events of VTE and their complications and potential death for hospitalized patients, in particular because their mobility is limited. The role networks consisted of circles (i.e. the *roles or team members*) and interactions between the circles. *Tasks* are either individual tasks and assigned to a specific role or team interactions and represented by an arrow between roles. *Tools and technologies* are represented by icons next to the tasks. A role network analysis visualizes the team and its interactions involved in a care process and can be complemented by other methods, such as quantification of network measures (Salwei et al., 2019).

Visualization has emerged as a key mechanism for understanding safety (Le Coze & Reiman, 2023). Several models of safety, such as the Swiss Cheese Model (Reason, 1997) and the sociotechnical model of Rasmussen (2000) that is the basis of Accimap, are represented by visual diagrams that tell us a lot about assumptions regarding safety (Waterson, 2023). Reason's model emphasizes latent and active factors leading to accidents, while Rasmussen's model focuses on interactions between multiple system levels. In the final chapter of their book on 'Visualizing Safety, An Exploration', Le Coze and Reiman (2023) bring up a series of questions on visualizing safety that could be adapted and used for improving our understanding of system interactions. For instance, participatory methods could be further developed to facilitate the involvement of various stakeholders in analyzing and visualizing a health care safety problem and identifying potential solutions.

Each SEIPS model is associated with a visualization that emphasizes specific features; see Figures 3.1–3.4. For instance, the SEIPS 3.0 model shows a series of interconnected work systems during the patient's journey. These different visualizations of SEIPS are often used in projects aimed at addressing a specific health care safety problem; we need to understand how these visualizations help participants understand system interactions. The visualizations may be boundary objects that help to establish common ground and facilitate collaboration between people coming from different disciplines and perspectives and working together on a project aimed at addressing a health care safety problem (Hall-Andersen & Broberg, 2014).

3.5 ENVIRONMENT AND CONTEXT

Environment and context are two terms that have different meanings and can be a source of confusion and misunderstanding. For instance, the terms work environment and context are often used to describe the work system. In this section, I clarify how the term environment is used in SEIPS. I then discuss the various interpretations of the term context that can be found in the literature on health care safety as well as the organizational design literature.

3.5.1 Environment in the SEIPS Family

In the SEIPS family, the term 'environment' is most often used to describe the *physical environment* and the *external environment*. The *physical environment* is about the physical aspects of the work system (Alvarado, 2012): where work is performed; the characteristics of the environment, such as lighting, noise and temperature; the characteristics of the workspace or workstation, such as distance to reach objects or supplies and adequacy of space to perform tasks; and the layout of the workplace, such as distance to supply room and design of patient room in relation to the workflow. There is evidence that the physical or built environment affects patient safety, such as HAIs (Zimring et al., 2013), medication safety and patient falls (Joseph et al., 2018), as well as health care worker safety (Hignett & Lu, 2010; Ulrich et al., 2008).

The *external environment* is outside of the boundaries of the work system and interacts with the work system; this is in line with systems thinking and the concepts of open system and system levels (Bertalanffy, 1950; Katz & Kahn, 1966); see discussion in Chapter 2 on Systems Thinking in Health Care Safety. The *external environment* includes the socioeconomic environment, community organizations, laws and regulations, demographics, political backdrop, cultural context, etc.

In the recent SEIPS 101 model (Holden & Carayon, 2021), we use the term environments to describe the *physical, socio-organizational and external environments*. We grouped the elements of the *physical environment, the organization and the external environment*; these elements are described separately in other SEIPS models. Whenever talking about the environment, it is, therefore, important to qualify what we mean by the environment and add an adjective or a descriptor, such as *physical environment* or *external environment*.

3.5.2 Context or Work System?

Similar to environment, context is a term that has vague and different meanings. In the health care quality and safety literature, the term context is often discussed, but rarely clearly defined. Dixon-Woods (2014) discusses the importance of considering the context when evaluating QI efforts in health care and proposes methodologies for considering context. She refers to characteristics of the context, such as those described by Shekelle et al. (2010), e.g. structural characteristics of the health care organization that is implementing the intervention, such as its quality and safety infrastructure. Shekelle et al. (2010) coined the concept of 'context-sensitive patient safety practices'. In addition to structural characteristics of the health care organization, they defined three other important characteristics of the context for patient safety: external factors (e.g. payment system); patient safety, teamwork and leadership at the unit level; and availability of implementation and management tools. In this conceptualization, the context is separate from the specific intervention (e.g. the patient safety practice): it influences the implementation, use and impact of the intervention. However, one can ask whether any patient safety practice is not context-sensitive. Separating a health care safety intervention from its context may be artificial: the intervention is part of the context, changes the context and is influenced by the context. A more fruitful approach is to define system levels and interactions between levels, including the 'context' or higher system levels and the changes that occur in the work system (i.e. lower system level) and interactions among system levels.

Context has been defined and conceptualized in many different ways in the literature on evaluation of health care quality and safety interventions (Greenhalgh & Manzano, 2022) and in the literature on organizational design and management (Johns, 2006; Rousseau & Fried, 2001). For instance, Johns (2006) describes different ways of thinking of the context, such as context as organizational opportunities and constraints that influence what is done in the local work system and context as the higher system level (e.g. the external environment) that influences what happens at a lower system level. Rousseau and Fried (2001) remind us that the term context comes from the Latin root: 'to knit together' or 'to make a connection' (page 1). Instead of talking about the context, they propose to focus on the action of contextualization, i.e. understanding what surrounds and influences a particular event, behavior or process. This can be done through a rich, detailed description of the care process, the work system and its external environment; this is the approach I take in this book.

In line with systems thinking, I agree with Greenhalgh and Manzano (2022) that there is not a single, stable context in which health care work occurs. Is the context the *external environment* of the work system, the *organization element* of the work system or the entire work system? Probably all of the above? In this book, I talk about the work system, the care process that is embedded in the work system and various system levels (e.g. local work system, health care organization, external environment). The term context is specifically used in the SEIPS 3.0 model to describe the socio-organizational level, i.e. the formal (e.g. hospital, primary care clinic, ED) and informal (e.g. home) care settings that interact with the local work system.

3.6 WHICH MODEL SHOULD ONE USE?

I am often asked the following question: 'Which SEIPS model should I use?' and the answer is… 'It depends!' The different members of the SEIPS family have been named to represent different versions: SEIPS, SEIPS 2.0, SEIPS 3.0 and SEIPS 101. They are different versions with unique features and emphasis. The most recent model is not necessarily better than the older models. Each model has unique characteristics and focus as shown in Table 3.1. For example, if your project is about understanding a care process and associated work system barriers and facilitators, you could use the initial version of SEIPS (Carayon, Hundt, Karsh, et al., 2006; Carayon, Wetterneck, Rivera-Rodriguez, et al., 2014). In many projects, the work system model (one of the core components of the SEIPS family) may be what you need (Carayon, 2009; Smith & Carayon-Sainfort, 1989).

You should use the model that fits your objective. In many projects, the original model published in 2006 (Carayon, Hundt, Karsh, et al., 2006) and its extension published in 2014 (Carayon, Wetterneck, Rivera-Rodriguez, et al., 2014) can provide the direction that you need; e.g. help with understanding what a 'system' is, ensuring that a wide range of system elements and characteristics is considered, analyzing how the work system elements are combined in a care process, etc. The SEIPS 2.0 model is particularly useful to distinguish different types of 'work' (Holden et al., 2013): not just work by health care workers but also patient work and the collaborative work between health care workers and patients. If a project focuses on the experience of patients, the SEIPS 3.0 model can be useful as it highlights the patient and their transitions between multiple care settings and work systems (Carayon, Wooldridge, et al., 2020). The SEIPS 101 approach provides practical tools and examples and is useful in applied improvement projects and for people with beginner's HFE knowledge (Holden & Carayon, 2021). So, use whatever SEIPS model fits your project and your objective; just make sure to think in systems!

3.7 KEY HFE FEATURES OF SEIPS FOR IMPROVING HEALTH CARE SAFETY

A landmark paper by the IEA Future of Ergonomics Committee identifies three fundamental characteristics of HFE (Dul et al., 2012):

1. HFE uses a systems approach,
2. HFE addresses multiple objectives for the person(s) and
3. HFE is design driven.

The key features of SEIPS are related to these three fundamental characteristics of HFE, i.e. systems approach, person-centered approach and a continuous cycle of design, redesign, improvement and learning; see Table 3.4.

TABLE 3.4
HFE Features of SEIPS

HFE Features	Application to SEIPS
Systems approach	Model of the work system with interacting elements Care process embedded in the work system Boundaries of the care process
Person-centered approach	Person at the center of the work system model Multiple outcomes that are interrelated Multiple sources of information about 'good' work systems; e.g. HFE design principles, literature on job/organizational design and job stress
Design, redesign, improvement and learning over time	Feedback loops of adaptation, improvement and learning Cycles of design and redesign Continuous design

3.7.1 SEIPS As a Systems Approach

As indicated in Chapter 2 and Table 2.1, the core concepts of systems thinking can be summarized as:

1. holistic orientation to systemic problems,
2. the recognition of complex network structures with multiple dynamic interactions among system elements and
3. the importance of multiple perspectives to address a problem or event.

SEIPS is aligned with these core concepts of systems thinking. SEIPS is foremost a systems approach that provides a structured means of analyzing the work system(s) in which a care process is embedded that influences health care safety. The work system is defined by its elements (see Table 3.2) and their dynamic interactions. The list of elements helps to ensure that all aspects of the work system are considered; it is a guide for understanding and analyzing the system, but it is not the end goal as interactions exist among system elements, which can be even more important than the elements themselves. The work system is a core component of SEIPS, but SEIPS is more than the work system; the work system is related to and influences the care process, which in turn has an impact on multiple outcomes. SEIPS also includes feedback loops of adaptation, improvement and learning.

3.7.1.1 System Boundaries

When analyzing work systems and care processes, a major question arises concerning the boundaries. What is included in the work system? What is part of the external environment? What is the beginning and end of the care process under study? Work systems and care processes involve various types of boundaries (Carayon, 2006):

- organizational boundaries: e.g. transition of care between hospital units, hospital discharge to home or long-term care,
- geographical boundaries: e.g. telehealth with patients and clinicians in different locations, virtual ICU with patients and clinicians in an ICU interacting with clinicians in a remote virtual ICU location,
- cultural boundaries: e.g. care of culturally diverse patient populations and
- temporal boundaries: e.g. patient care across multiple work shifts of hospital staff.

Here, we focus on organizational boundaries of the work system and care process. Looking at the patient's experience and their interactions with health care, we aim to identify the organizations and care settings that the patient interacts with. We recognize that organizational boundaries can be related to other types of boundaries. For instance, crossing organizational boundaries between a hospital and a long-term care facility may also involve geographical (different locations for the hospital and the long-term care facility), cultural (different professional cultures and mental models) and temporal (different work schedules in the hospital and long-term care facility) boundaries.

A study of care transitions of older adults between the ED and long-term care by Werner et al. (2021) shows interaction between organizational and cultural boundaries. Staff in the ED from one hospital (e.g. physicians, nurses, case managers) and in four skilled nursing facilities (SNFs) (e.g. administrators, nurses, coordinators) were interviewed about care transitions of older adults between the ED and the SNF. The thematic qualitative data analysis identified several themes that represent ED and SNF staff's misaligned mental models regarding the care transition of older adults between the ED and the SNF. For instance, the ED staff and the SNF staff had different views on who was responsible for communicating patient-related information and what was the best way of communicating that information. The ED and SNF are two organizations that are part of the care transition, but the ED and SNF staff have different mental models of the care transitions and aspects of the work system that help or hinder the care transition and ultimately the older adult's safety. This example shows that enlarging the system's boundaries, i.e. considering both the ED and the SNF, is important to address safety for older adults transitioning between the two care settings. Understanding different perspectives on a specific health care safety problem is critical to define the system's boundaries.

3.7.1.2 Boundaries Between Work Systems

In conducting a SEIPS analysis of a care process, it is important to identify all of the work systems involved and the boundaries and interactions between the work systems; this helps to determine the boundaries of the problem that will be the focus of the improvement project. A study of discharge of older adults from the ED to their home identified six different work systems involved in the care transition (Werner, Rutkowski, et al., 2020): the patient's home, the family, the community, the primary care clinic, the community pharmacy and the ED. Understanding interactions within each work system and between work systems is key to improving health care safety as the patient transitions from the hospital to home.

As multiple work systems interact with each other, barriers and facilitators within and across work systems play an important role in boundary-spanning activities. For instance, ambiguity or confusion about who is responsible for medication management may occur when older adults transition between the hospital and their home (Werner et al., 2017). The older adult recovering at home after a hospitalization may wait for the visit of a home health nurse to ask questions about medications; this could have a negative impact on medication safety if there is a delay in adapting the medication regimen. Boundaries between work systems and around the system and care process under analysis are, therefore, critical in health care safety. SEIPS can help in analyzing these complex systems, but a major question is how and where to set the system's boundaries to deal with a specific health care safety problem.

3.7.1.3 Defining System Boundaries

How do we determine the boundaries of the system under study? When using a systems approach such as SEIPS, it is important to agree on the boundaries of the system(s) that will be the target of analysis in order to improve health care safety for that system(s) and the people involved. Because the work system is complex and open (i.e. it interacts with an external environment) and multiple work systems are inter-connected in a care process, the boundaries of the system(s) and the care process can be fuzzy and, therefore, challenging to define in any improvement project (Eason, 2014). As stated by Wilson (2014), 'There are no real rules for deciding where to define the boundaries for the system of interest, other than being clear and sensible enough to acknowledge that the setting of the boundary must be useful for the purposes of the analysis' (page 7).

Defining the system's boundaries is necessary to proceed with the analysis and improvement, and the decision about boundaries is a balancing act. Boundaries should be wide enough to account for the major factors that could potentially be addressed in the work system redesign; they should not be so wide that the improvement project becomes too big and potentially not feasible or manageable; nor should they be so narrow that key factors are not included in the analysis (Siemieniuch & Sinclair, 2006). In addition, as advocated by many authors (Greenhalgh & Papoutsi, 2018; Wilson, 2014), we need to be flexible in defining the boundaries. We may change the boundaries, making them wider or more narrow, as the project evolves and we learn more about what contributes to health care safety and how to design and implement work system interventions.

Defining the boundaries of the system under consideration should be a collaborative process that involves key stakeholders (Klein, 2014). As different people (and roles) have different perspectives on a health care safety problem (Werner et al., 2021), the discussion about system boundaries is likely to involve values about what is important and what can (or should be) changed. Input from stakeholders regarding the care process and its boundaries may be a useful initial step. In a SEIPS study of care transitions between the OR and the ICU, Wooldridge et al. (2022) asked OR and ICU staff (e.g. surgeon, anesthesiologist, critical care physician, OR and ICU nurses) to identify the beginning and end of the OR–ICU care transition. A SEIPS-based process map of the OR–ICU care transition was developed and included the following major steps:

- transition preparation: e.g. end of surgery and OR preparation to move patient, ICU preparation to receive patient,
- transition: e.g. verbal and/or face-to-face handoffs between the OR and ICU staff, moving the patient to the ICU and
- transition follow-up: e.g. documentation, continuing patient care.

OR and ICU health care professionals had different views on the beginning and end of the care transitions; the variation was explained by their role (e.g. physician, nurse) as well as their location (i.e. OR, ICU). Identifying different views and integrating multiple perspectives are critical in defining (or redefining) the boundaries of the system and care process under study; this is further discussed in Section 4.8.4 of Chapter 4 on Human-Centered Design and SEIPS in Action.

3.7.2 Person-Centered Approach

The SEIPS family is anchored in the HFE discipline; therefore, the person is always at the center of the work system. The system and associated care process should be designed to help the person and produce multiple positive outcomes. Health care safety is about patient safety and health care worker safety; therefore, we need to look at various approaches and bodies of literature to identify what makes a 'good' work system that produces positive outcomes for patients and health care workers.

3.7.2.1 Designing Patient-Centered Care Processes

We need to design work systems and care processes that are centered around the patient. Numerous reviews of the literature help to understand the attributes or dimensions of patient-centered care (Castro et al., 2016; Eklund et al., 2019; Scholl et al., 2014), the factors that help or hinder patient-centered care (Castro et al., 2016; Grover et al., 2022), and the relationship of patient-centered care to other concepts, such as patient empowerment (Castro et al., 2016), shared decision-making (Stiggelbout et al., 2015) and person-centered care (Eklund et al., 2019). From a SEIPS perspective, we need to understand the work system in which patient-centered care occurs and the characteristics of the work system that hinder or facilitate patient-centered care. This is why models that expand the SEIPS family and conceptualize patient work are critical (Holden, Schubert, & Mickelson, 2015; Holden et al., 2017; Werner, Ponnala, et al., 2020).

Patients (and care partners) do work; i.e. they visit their primary care physician, they take medications, they measure their blood pressure, etc. Analyzing patient work and the various systems in which patient work occurs is critical to improving it and removing barriers to patient-centered care. The literature on patient ergonomics provides numerous principles for designing patient-centered work systems and improving health care safety (Holden & Valdez, 2021; Valdez & Holden, 2021).

3.7.2.2 Models and Design Principles for Health Care Work Systems

From a health care worker safety perspective, we know a lot about the characteristics of work systems that produce positive worker outcomes. Over the years, the literatures on HFE, job/organizational design and job stress have developed numerous models and principles for 'good' work systems and their elements. The job stress literature has produced multiple models that define psychosocial work factors contributing to job stress, such as the Job Strain model (Karasek & Theorell, 1990), the Job Demands–Job Resources model (Bakker & Demerouti, 2017) or the Effort–Reward Imbalance model (Siegrist, 2017). As highlighted in Chapter 1 on Health Care Safety, the National Academies' report on clinician burnout provides a list of work system factors that are organized as job demands and job resources (NASEM, 2019); see Table 3.5. The list of job demands and job resources can be translated into design principles for 'good' work systems: the job demands should be reduced, eliminated or mitigated, and the job resources should be enhanced.

The job/organizational design literature has developed a range of models about what makes a 'good' job, such as the Job Characteristics Theory (Hackman & Oldham, 1976) and the recent comprehensive SMART work design model (Centre for Transformative Work Design, 2025). As SEIPS is anchored in the STS Theory, I highlight the STS principles of job design (Davis, 1977; Trist, 1981); see Table 3.5.

Several job demands and resources described in the National Academies' report on clinician burnout (NASEM, 2019) are similar to the STS job design principles,

TABLE 3.5
Work System Factors of Importance to Psychological Aspects of Health Care Worker Safety

Work System Factors Contributing to Clinician Burnout and Professional Well-Being (NASEM, 2019)

Job demands:	Job resources:
• Excessive workload, unmanageable work schedules and inadequate staffing • Administrative burden • Workflow, interruptions and distractions • Inadequate technology usability • Time pressure and encroachment on personal time • Moral distress • Patient factors	• Meaning and purpose in work • Organizational culture • Alignment of values and expectations • Job control, flexibility and autonomy • Rewards • Professional relationships and social support • Work–life integration

STS Theory Principles of Job Design (Davis, 1977; Trist, 1981)

- Job content reasonably demanding with variety
- Opportunity for learning
- Autonomy and opportunity for decision-making and discretion
- Social support in the workplace
- Recognition of one's performance
- Meaningful social contribution
- Desirable job future

such as meaningful work and job content, and job control and autonomy. While many principles in the job/organizational design and job stress literatures have been developed for paid jobs, they may also be adapted and applied to improve the work of patients and care partners, as suggested by Werner (2025).

3.7.2.3 HFE Design Principles

The HFE literature has developed numerous principles about work system design, such as the usability of health information technologies, physical design of workspaces and the design of teamwork. These principles can be applied to improve the design of work systems for both patients and workers. Inadequate usability of technologies is mentioned in the list of job demands contributing to clinician burnout (see Table 3.5). The HFE literature provides specific guidance on how to design usable *technologies*, such as clinician-oriented medication alert systems (Marcilly et al., 2023) and patient-oriented discharge summaries (Tremoulet et al., 2018). The *physical environment* is a work system element that is important for both patient safety and health care worker safety. Specific design principles for the *physical environment* include (Carayon, Alvarado, & Hundt, 2007a):

- minimizing perception time (e.g. visual discrimination with the use of appropriate size and color),
- minimizing decision time (e.g. spatial compatibility),

- minimizing manipulation time (e.g. ease of manipulation),
- optimizing opportunity for movement (e.g. ease of access to equipment) and
- minimizing the need for human strength (e.g. patient beds with built-in scales).

The *organization* element of the work system includes teamwork. The TeamSTEPPS program, developed by the US Department of Defense (DoD) and AHRQ, provides guidance on what makes for good teamwork in health care (King et al., 2008); for more information, see: https://www.ahrq.gov/teamstepps-program/index.html.

All of these HFE design principles need to be integrated into health care safety improvement projects; see Chapter 4 for examples. It is not possible to list all of the HFE design principles as this knowledge is vast and deep; see various handbooks for additional information on HFE (Carayon, 2012; Salvendy & Karwoski, 2021) and the recent book by Sujan et al. (2025). In a chapter published in the first edition of my 'Handbook of Human Factors and Ergonomics in Health Care and Patient Safety' (Carayon, 2007), we show how various models and concepts of the literature on job/organizational design, job stress, and HFE are related to work system elements (Carayon, Alvarado, & Hundt, 2007b). This provides guidance on where to find information on design principles for 'good' work systems.

As the work system is a complex network of various elements, we need to look at how to design the elements as well as their interactions. This is where the Balance Theory comes in handy, as it provides 'meta' guidelines on how to design the entire work system. We need to ensure that the individual elements are well-designed and that negative aspects of the work system are addressed; this requires a realistic and pragmatic approach where the compensatory design and overall system balance principles are considered. See Section 3.2.2 for additional discussion of the Balance Theory.

3.7.3 Design, Redesign, Improvement and Learning Over Time

The SEIPS family integrates time and change in both explicit and implicit ways. A core component of the SEIPS family is the feedback loops between work systems, processes and outcomes; this is one explicit way that time and change are represented in SEIPS. The feedback loops are important as they represent continuous cycles of design and redesign and the mechanisms of adaptation, improvement and learning over time. These cycles of design and redesign could almost be endless as change is implemented, people react to the change, lessons learned emerge from the implementation of the change, this information is then fed back to the work system, and additional changes are implemented, and so on. All of this occurs in a dynamic external environment that introduces new disturbances, constraints and opportunities. In those dynamic and continuous cycles of design and redesign, it is important to foster individual and organizational learning.

Because the work system is dynamic, it changes and people react to it and adapt it. The learning and adaptability of people is key to consider in the cycles of design and redesign (Eason, 2014). As discussed in Section 3.4.4, adaptation is critical in

SEIPS. Understanding how people adapt to the work system can provide important information that feeds into individual and organizational learning and help to continuously improve health care safety. Working around challenges or barriers in the work system has been observed in many health care processes, such as care coordination for chronically ill patients (Carayon et al., 2019), cardiac surgery (Gurses, Kim, et al., 2012) and home care for children with medical complexity (Barton et al., 2021). Work-arounds are strategies that help people perform their work. Work-arounds and other strategies are reactions to the work system or ways of dealing with the work system.

The work system can be used to describe work-arounds. In a study of physicians, nurses and clerks in an academic Dutch hospital, Blijleven et al. (2017) identified 15 EHR-related workarounds associated with four of the five work system elements. For example, physicians and nurses valued face-to-face interaction and, therefore, wrote down information during their encounter with patients rather than using the computer (*technology element of the work system*). Could this work-around be addressed by EHR redesign (*technology element*) or by a change in the workflow (*task element*)? The researchers suggested different solutions to address the work-arounds depending on their association with work system elements.

Work-arounds are often viewed negatively and as creating potential safety issues. One could be concerned about work-arounds, in particular those linked to safety standards and policies. However, a recent systematic review shows an ambiguous impact of work-arounds on safety (Clark et al., 2025): some studies showed a negative impact of work-arounds on safety, while other studies showed a positive impact. This calls for attention to the design of safety programs; they should be designed to support the work of people and not hinder their ability to do work. If safety programs are not well-designed, they may produce hazards and actually defeat the purpose of improving health care safety; this is discussed in Section 1.3 of Chapter 1 on linkages between patient safety and worker safety.

<u>One should not wait for a major health care safety problem to arise before taking action and engaging in a cycle of design and redesign with the goal of continually improving health care safety and learning throughout the process and over time.</u> Health care organizations need structures and processes for capturing early warning and weak signals before adverse events and disasters occur (Macrae, 2014); this could involve the voice of patients and families (Wiig et al., 2024). This is a challenging endeavor as health care organizations need to balance 'what to pay attention to, what to investigate, what to act on—and what to ignore' (Macrae, 2014, p. 444). HFE provides the tools, methods and approaches to support some of these activities. Because HFE is design driven, it has developed numerous methods and approaches for analyzing and improving work systems. HCD and its linkages to HFE and SEIPS are the topic of Chapter 4.

4 Human-Centered Design for Health Care Safety

SEIPS in Action

Human-centered design (HCD) is increasingly used and emphasized in health care in order to address various problems (Göttgens & Oertelt-Prigione, 2021; Melles et al., 2021), including patient safety and health care worker well-being (Benishek et al., 2023). HCD can take different forms, including user-centered design (UCD), design thinking, collaborative design and participatory design (Sánchez de la Guía et al., 2017; Sanders & Stappers, 2014). There is sometimes confusion and misunderstanding about HCD (Proctor et al., 2021). HCD is sometimes equated to a specific approach, such as design thinking or UCD (Göttgens & Oertelt-Prigione, 2021; Rodriguez et al., 2023). HCD is much more than a specific approach. HCD includes a range of design processes with various activities and methods and has specific characteristics of human-centeredness, such as empathy and participation.

I use the term HCD as the umbrella concept to describe different approaches. In this chapter, I first review what HCD is and different ways of thinking of HCD: HCD as a process, HCD methods and characteristics of HCD. As I am anchored in the HFE discipline, I conceptualize HCD as the 'how to' of HFE. Adapting the ISO model of HCD (International Organization for Standardization (ISO), 2010), I then describe various steps of the HCD process and show how SEIPS is integrated in HCD. A major part of this chapter is dedicated to specific applications of SEIPS in HCD, including analysis of work systems and care processes, design, implementation and evaluation of work system interventions, and analysis of adverse events. In Chapter 5, I describe three projects in which SEIPS was used in different steps of the HCD process; these case studies provide additional information on how SEIPS can be used in practice.

4.1 INTRODUCTION: HCD AND HFE

There has been a proliferation of papers about HCD and its applications to improving health care safety. It is important to recognize that HCD and HFE are interrelated; they could be considered as two sides of the same coin. HCD is the 'how to' of HFE, i.e. the process by which HFE knowledge (e.g. HFE design principles, methods and models; see Chapter 3 on what makes a 'good' work system) is applied. Indeed, the ISO definition of HCD shows how HFE is integrated in the HCD process (International Organization for Standardization (ISO), 2010):

> "Human-centred design is an approach to interactive systems development that aims to make systems usable and useful by focusing on the users, their needs and requirements,

DOI: 10.1201/9781003480310-4

> and by applying human factors/ergonomics, and usability knowledge and techniques. This approach enhances effectiveness and efficiency, improves human well-being, user satisfaction, accessibility and sustainability; and counteracts possible adverse effects of use on human health, safety and performance".
>
> (page vi)

The ISO definition of HCD is clearly aligned with the HFE discipline as HCD uses HFE knowledge and aims to produce interventions and innovations that improve multiple outcomes of usability, performance, well-being and safety, i.e. the multiple objectives of different stakeholders (Dul et al., 2012).

HCD without HFE is like an empty shell: how do we know what makes for a 'good' design? The HFE discipline has developed a wide range of knowledge about design principles, such as usability heuristics, physical design of devices and equipment, balancing workload, design of team training programs, enhancing safety climate and culture, etc. (Sujan et al., 2025). These HFE design principles should be integrated into HCD to produce a 'good' solution or intervention. Knowledge about 'good' work system design may not always come in specific design principles, but in general guidance. In Chapter 3, I describe various characteristics of the work system that are important for health care safety. For instance, workload should be reasonable, i.e. neither too high (overload) nor too low (underload), or workers should have the opportunity for autonomy and control of facets of their job. These examples of 'good' work system characteristics do not provide specific guidance for how to redesign care processes; the principles need to be used, adopted and adapted in specific improvement projects.

The strong interest in applying HCD in health care has led to a proliferation of papers and commentaries, sometimes with limited knowledge and understanding of HCD. There is deep knowledge in engineering psychology and HFE on HCD (Proctor et al., 2021) that has been developed since the 1950s. Not understanding this historical perspective can lead to 'reinventing the wheel' and losing precious time and resources for improving health care safety. This is why involving HFE professionals in health care safety improvement projects is important.

Co-design or collaborative design is one of the numerous approaches of HCD. Recently, we have seen an increased number of 'co-'approaches in health care: e.g. co-production, co-design and co-creation (Grindell et al., 2022). Co-production and co-design are increasingly discussed together (Batalden et al., 2016; Masterson et al., 2022; Robert et al., 2022), with co-production representing the way patient care is or should be designed (e.g. patient-centered care, shared decision-making) and co-design being the process for involving patients in designing patient-centered care processes. Co-design is the 'how to' and co-production is the outcome of the design process. Unfortunately, as these 'co-'approaches are being discussed, little or no reference is made to the deep knowledge in HFE and HCD (Grindell et al., 2022). This is another demonstration of the need to better disseminate and integrate HFE knowledge in health care; this is one of the goals of this book.

HFE and its correlate, HCD, are core to redesigning complex health care work systems and improving health care safety. As discussed in Chapter 2, the systems

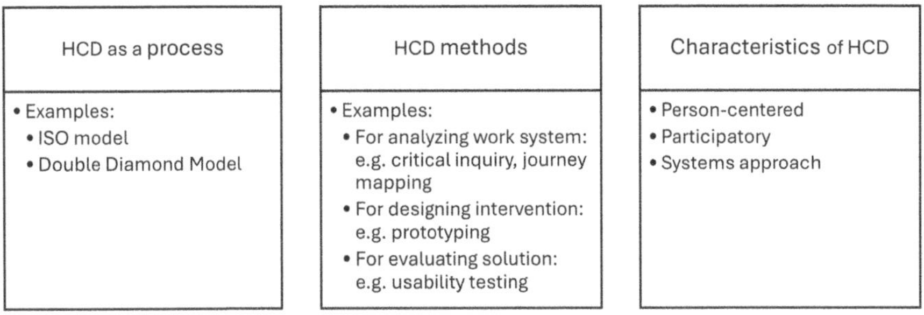

FIGURE 4.1 What is Human-Centered Design (HCD)?

thinking literature tells us that intervening in complex systems is difficult. We need to think of improving complex health care systems not as the search for a single solution, but as a process with numerous opportunities for design, redesign, learning and continuous improvement. HCD provides such opportunities for action, feedback and learning, and, therefore, can support the development of individual and organizational capabilities for long-term continuous improvement of health care safety (Fraser & Greenhalgh, 2001).

4.2 DIFFERENT WAYS OF THINKING OF HCD

HCD includes a range of approaches with specific activities and methods; see Figure 4.1. HCD is often conceptualized as a process with activities that unfold over time. In this chapter, I describe two process models of HCD: the ISO model (International Organization for Standardization (ISO), 2010) and the Double Diamond Model (Design Council, 2025). HCD is also described as a set of methods, e.g. methods for analyzing work systems (e.g. critical inquiry, journey mapping), for designing interventions (e.g. prototype), for evaluating solutions (e.g. usability testing), etc. Given the close relation between HCD and HFE, it is important to recognize that the HCD methods need to be complemented by systems thinking as it applies to work (e.g. SEIPS) and criteria for defining what makes for 'good' work (i.e. HFE design principles). Finally, HCD has three core characteristics: person-centered, participatory and systems approach (Melles et al., 2021).

4.2.1 HCD as a Process

HCD is often described as a process that unfolds over time and consists of various activities linked to each other. I describe two process models of HCD that are often used in health care safety: the ISO model (International Organization for Standardization (ISO), 2010) and the Double Diamond Model (Design Council, 2025).

4.2.1.1 ISO Model of HCD

According to ISO (2010), the HCD process consists of four inter-related activities:

1. Understanding and specifying the context of use, i.e. the work system in which the new technology, work procedure or any other intervention will be used.
2. Specifying the user requirements, including the needs of users and the requirements of the intervention to be developed. The requirements are derived from the analysis of the work system (step 1) as well as HFE knowledge, standards and guidelines (or what I call 'HFE design principles').
3. Producing design solutions, such as prototypes, simulations or mock-ups. This step also includes early evaluation that can help to select and/or improve the solution. This is a highly iterative part of the HCD process with potentially multiple feedback loops to the previous steps.
4. Evaluating the design, including evaluating early prototypes, conducting experimental evaluation and evaluating the solution in the field. Evaluation methods are diverse and include direct participation of the users (e.g. scenario-based user testing) and input from experts (e.g. heuristic usability evaluation).

These four activities are interdependent as the output of one activity is used as input for another activity. In addition, the ISO model clearly indicates that the activities are not organized in a linear process. For instance, it is possible to go back to analyzing the work system (step 1) after evaluating a prototype (step 4). The design team may recognize that they lack information on specific aspects of the work system and, therefore, go back to their initial phase of understanding the context of use. In addition, the ISO model emphasizes 'long-term monitoring' or learning, i.e. the feedback loops in SEIPS. This involves follow-up evaluation of the intervention over a long period of time, which could help to identify unanticipated consequences and lead to revisions of the initial intervention; a process that we call 'continuous design' (Carayon & Salwei, 2021).

Let us look at one application of the ISO model in health care safety. Emergency departments (EDs) around the world are facing overcrowding issues, which can lead to patient safety issues; e.g. urgent cases being delayed or patients leaving without being seen by an ED professional (Derlet, 2002; Perry et al., 2012). Patient triage upon arrival to the ED is, therefore, critical to ensure patients with critical conditions are seen in a timely manner. Schiro et al. (2020) developed a patient prioritization tool for a pediatric ED in a French hospital. They used the ISO model of HCD and performed the following activities:

- Work system analysis of the pediatric ED with a focus on the triage process, e.g. understanding clinicians' decision-making process for prioritizing patients, and information needs for the triage process. This analysis included the collection of observation data (structured observations performed by an HFE professional) and interviews of ED clinicians.

- Requirements for the patient prioritization tool, which were based on data from the work system analysis and were further refined with focus groups of clinicians.
- Design of a mock-up of the patient prioritization tool.
- Evaluation of the mock-up in a four-phase user testing process with seven ED nurses and five ED physicians. The user testing process alternated testing and training sessions. The results demonstrated high usability of and satisfaction with the tool. Some usability issues were identified and addressed in the next iteration of the tool's design.

4.2.1.2 Double Diamond Model

The Double Diamond Model (Design Council, 2025) was developed by the UK Design Council in 2003 and has been applied by numerous designers around the world in multiple domains, including health care. The Double Diamond Model consists of four activities distributed in two main groups:

1. Problem space with activities of discovering and defining, which help and contribute to defining and clarifying the problem.
2. Solution space with activities of designing and validating, which lead to the implementation of a solution.

Within each of the two spaces, i.e. problem space and solution space, divergent thinking and convergent thinking occur. For instance, in the problem space, the initial activity of discovering is broad, uses multiple methods, and explores various problems; this divergent phase leads to the convergent phase of defining the specific problem to be addressed in the solution space. A recent iteration of the model includes feedback loops within each of the two spaces or diamonds, as well as between the spaces or diamonds and back to the initial phase (Design Council, 2025).

Let us look at one application of the Double Diamond Model in health care safety. Providing up-to-date and critical information to hospitalized patients is not only important for their experience and satisfaction, but it also supports timely communication between patients and the hospital staff. Patients need more and better information, which can help them prepare for interacting with health care professionals, e.g. communication during bedside rounds. Yoo et al. (2015) designed a 'smart bedside station', i.e. a touch-screen terminal located at the bedside that provides patients with a range of information and convenient access to hospital services. They used the Double Diamond Model to design the technology as follows:

- Discover: observation of nurses interacting with patients, role playing of researchers as patients and care partners, and contextual inquiry of patients and care partners in various hospital units.
- Define: mapping the patient's journey during the hospitalization and an affinity diagram of issues and unmet needs. Both methods used data collected during the discovery phase.

- Develop: collaborative workshop with researchers, physicians, nurses and HFE professionals, which produced a range of ideas that were later organized into clusters. Prototypes were then developed and tested.
- Deliver: final outputs with framework sketch, visual scenario and interface guidelines.

These examples of the application of the ISO model and the Double Diamond Model show how various HCD processes are implemented to produce beneficial solutions for improving health care safety. The ISO model and the Double Diamond Model involve multiple steps of analyzing, defining and refining the solution and evaluating the solution. There can be multiple iterations of each step as well as feedback loops between the steps, e.g. going back to the analysis after realizing gaps in knowledge about the work system. As shown in the examples described in this section, HCD includes a range of methods for collecting, analyzing, using and sharing data.

4.2.2 HCD Methods

The HCD methods are multifarious. Multiple books and resources provide details on HCD methods, such as HFE books (e.g. 'Handbook of Human Factors and Ergonomics Methods' (Stanton et al., 2004) or the 'Handbook of Human Factors and Ergonomics' (Salvendy & Karwoski, 2021)), and books written by designers (e.g. 'The Convivial Toolbox' by Sanders and Stappers (2013) or the 'Delft Design Guide' (van Boeijen et al., 2013)). Data collection for HCD relies on a range of methods, including observation, interview, focus group and survey. The data can then be analyzed with various techniques, such as journey mapping and use case scenarios. The various data, outputs and artifacts are used in collaborative design processes, such as participatory design workshops.

The narrative review by Göttgens and Oertelt-Prigione (2021) produced a long list of qualitative and quantitative data collection methods, which are used at all stages of the HCD process, and design methods for each of the HCD phases. The methods are organized according to the four steps of the ISO model: understanding the work system, specifying the problem and user needs, generating ideas and designing solutions and evaluating solutions. Some design methods can be used at multiple stages of the HCD process. For instance, personas can help in understanding the context of use or work system and for specifying the problem; expert-based heuristic evaluation and think-aloud techniques with potential or actual users are used when designing solutions or evaluating solutions. Melles et al. (2021) described a selected set of HCD methods for each step of the Double Diamond Model:

- Discover: observations, interviews, generative techniques
- Define: stakeholder mapping, patient journey mapping
- Design: brainstorming, co-creation session or design charette
- Validate: interaction prototyping.

Specific examples of HCD methods used to improve health care safety are described in subsequent sections of this chapter as well as in Chapter 5 with case studies that focus on SEIPS-based methods.

4.2.3 Core Characteristics of HCD

HCD is an iterative process of activities that uses multiple, diverse methods. In addition to the 'technical' aspects of HCD, we need to consider its underlying values and characteristics. As HCD is the 'how-to' of HFE, it puts people at the center of the process, and it is anchored in systems thinking. Melles et al. (2021) defined three key characteristics of HCD:

1. HCD emphasizes people, i.e. understanding their needs and designing a solution that meets their needs and constraints.
2. HCD is based on the participation of multiple stakeholders. Participation occurs early and is continuous throughout the HCD process.
3. HCD integrates a systems approach.

4.2.3.1 Empathy in HCD

HCD involves a range of activities that directly involve people, such as observation or interview. These methods are aimed at understanding people's needs, their constraints, their perception of barriers and facilitators, the impact of these barriers and facilitators on patient safety and worker safety and their ideas for improving the work systems, care processes and outcomes. As health care is about people caring for people who experience a disease, injury, discomfort, sickness, life-and-death situations, etc., we need to conduct HCD work with empathy. As described by Tim Brown (2009), the chair of IDEO, an international design company, we could spend a lot of time gathering various data and notes in order to understand users and their work system, but this could not be done without connecting with the people that we aim to help; therefore, empathy is critical.

Tim Brown (2009) describes empathy as 'standing in the shoes (or lying on the gurneys) of others' (page 49). This can be done, for instance, with HFE analysts playing or simulating the role of patients or workers in care processes. Engaging participants (e.g. patients, health care workers) in analyzing their work systems or in role-playing potential solutions supports the development of meaningful relationships and trust between the people whose work systems we are attempting to improve and the HFE analysts. Empathy is critical; it is about HFE professionals demonstrating their human values (Suri, 2001).

4.2.3.2 Participation in HCD

Worker participation is a well-accepted principle in organizational design. Participating in decision-making provides workers an opportunity to influence what is done, what is implemented and how it is implemented (Wilson, 1995); this has multiple benefits, such as solution acceptance, effective implementation, increased fit of solution with the work system, etc. But what is participation? If a health care professional fills out a questionnaire survey about the pluses and minuses in their job, is that participation? The worker is participating as s/he provides information that is used in the HCD process, but this participation is limited. In line with Wilson (1995), I conceptualize participation as addressing significant issues in the work system with impactful activities during the HCD process.

Participatory practices described in the HFE literature vary along numerous dimensions, such as the content of the participatory project (i.e. what is designed), who participates and what decisions participants can make (Haines et al., 2002). Already in the 1990s, Ken Eason (1995) called our attention to different HFE approaches to participation in HCD; he contrasted 'design for users' and 'design by users'. Design for users would typically involve the application of HFE technical knowledge with users being limited to 'passive participants' (e.g. filling out a questionnaire, being observed). Design by users is what is typically involved in participatory ergonomics: the users influence the design process, the design solutions and their implementation. Whereas participatory ergonomics has primarily focused on addressing physical aspects of the work system and the issue of WRMDs (Hansen et al., 2024), a few applications have addressed health care safety, such as the design of family-centered rounds in a pediatric hospital (see case study #1 in Chapter 5) (Xie, Carayon, Cox, et al., 2015).

Participation is multifarious. Participation can be direct or indirect. Direct participation may involve a small group of health care professionals or patients actively engaged in design workshops where design solutions are being created. When a person is designated as the representative or the delegate of a larger group to participate in a design workshop, the larger group is indirectly participating in the HCD process via their representative. Participation should include people at the sharp end (e.g. patients, physicians, nurses) as well as stakeholders at the blunt end, such as organizational leaders (e.g. top managers, unit directors), institutional stakeholders (e.g. union representatives) and operational staff (e.g. IT professionals). See different examples of how this is done in the case studies described in Chapter 5. Participatory practices can be challenging to implement in HCD. However, it is important to remember that participation should be genuine; it should not be used as a superficial method to give patients or workers the illusion of control. Creating opportunities for people to express their opinions in a safe environment is a marker of genuine participation (Østergaard et al., 2018).

Kushniruk and Nohr (2016) proposed a categorization of HCD according to the type and level of participation; they distinguished between user-centered design (UCD), co-design and participatory design. They discussed these three HCD approaches in the context of health IT; however, these can be more broadly applied to any work system redesign aimed at improving health care safety. UCD was proposed by Norman and Draper (1986). UCD often involves usability-related methods, such as heuristic evaluation and usability testing; it is based on user input in design with primarily indirect participation. It is called 'design for users' by Eason (1995). In collaborative design, users and designers work closely together and cooperate to develop and implement a solution (Détienne, 2006). Co-design is what Eason (1995) calls 'design with users'. Finally, in the Scandinavian concept of participatory design, users are in charge of developing solutions; designers accompany the design process (Bratteteig & Wagner, 2014). Participatory design aims at supporting human endeavor and increasing autonomy and decision-making power for the users (Bratteteig & Wagner, 2016). For further description of these three HCD approaches and examples, see Kushniruk and Nohr (2016) and Carayon et al. (2025).

4.2.3.3 Systems Approach in HCD

In Chapter 2 on Systems Thinking, I highlight the concept of emergence: i.e. the design of a system emerges as an output of multiple interactions over time. Emergent systems may produce good or bad safety outcomes. In line with Rouse (2008), I believe that we can engage in HCD activities to improve the design of complex health care systems and, therefore, health care safety: 'As much as possible, complex systems should be designed and not just emerge' (Rouse, 2008, p. 21). Even when the focus of HCD is one part of the system, e.g. designing a clinical decision support (CDS) *technology* for supporting the work of physicians in the diagnostic process of a clinical problem such as PE, we are not designing a *technology* in isolation. The *technology* interacts with many other elements of the work system, which need to be considered in the HCD process: for instance, where the CDS will be used (i.e. *the work system element of physical environment*), how the CDS will be used (i.e. *the work system elements of task and organization*), who will use the CDS (i.e. *the person element*), etc. Therefore, the HCD process should integrate a systems approach. In Section 4.3, I describe how the systems approach of SEIPS is integrated with HCD.

4.3 HOW DOES SEIPS INTEGRATE WITH HCD?

I think about HCD as the 'how-to' of HFE. HCD is a process with activities, methods and characteristics that support the implementation of HFE knowledge and the objectives of HFE (i.e. improving health care safety). As the systems approach is a key characteristic of HCD (Melles et al., 2021), I propose that SEIPS should be fully integrated in HCD.

HCD has been described as a process with various steps; SEIPS can be integrated into all of the steps of the HCD process (International Organization for Standardization (ISO), 2010):

- understanding and specifying the context of use or work system,
- specifying user requirements,
- producing design solutions and
- evaluating the design.

Analyzing the so-called 'context of use' can be supported by the SEIPS model as the context of use can be represented by the work system and the care process. In subsequent sections, I describe various SEIPS-based methods for analyzing jobs or roles and care processes.

User requirements are specified using data produced from the initial analysis. The user requirements can be enriched by using SEIPS: user requirements should address all work system elements. User requirements should also use HFE knowledge about what determines a 'good' work system, i.e. HFE design principles.

At the stage of producing design solutions, prototypes may be developed. They can be physical prototypes of a redesigned physical space (e.g. a patient room), informational prototypes of a health IT (e.g. CDS), or process prototypes (e.g. simulation games (Vartiainen & Ruohomaki, 1994)). Whatever form the prototypes of design solutions take, they are all embedded in a work system. Therefore, I recommend the use of SEIPS-based description of the design solutions that addresses all of the work system elements.

Evaluating the design solution can be driven by the SEIPS model. Such a SEIPS-based evaluation examines the impact of the intervention on the work system, the care process and multiple outcomes of health care safety. SEIPS-based evaluation goes beyond typical health care intervention evaluations that tend to focus on measures of care process and outcomes. In addition to evaluating the impact of the intervention on process and outcome measures, we need to understand how the intervention fits (or does not fit) with the rest of the work system by identifying its work system barriers and facilitators.

SEIPS should be integrated with the HCD process along with other HFE models, concepts and methods. For instance, if the HCD process focuses on developing a health IT, usability is an important HFE concept to consider and multiple methods for improving and assessing usability should be incorporated in HCD. In this book, I focus on SEIPS while integrating other bodies of knowledge, in particular in the HFE discipline, that make important contributions to designing and implementing interventions and improving health care safety.

4.4 SEIPS FOR ANALYSIS

SEIPS has been used extensively to describe and analyze work systems and care processes. As explained above, any care process (or more broadly any process in health care) can be described using the work system elements: a process is a series of *tasks* performed by various *individuals* using *tools and technologies*; these activities occur in *physical and organizational environments*. This rich description of a care process can be contrasted with the concept of 'process' in traditional QI, which is often described as a linear sequence of tasks, but rarely discussed in light of the overall context or work system (Bowie & Vosper, 2024). SEIPS is frequently used to identify what works and what does not work in a care process, i.e. work system barriers and facilitators.

BOX – SEIPS ANALYSIS

Various models of the SEIPS family have been used to analyze work systems and care processes with the goal of improving health care safety. There is not one single SEIPS method. Multiple data collection methods are used, such as observation, interview and survey. In a SEIPS analysis, both qualitative and quantitative data are often collected. Multiple methods of data collection are combined, and, at the stage of analysis, data can be mixed to provide a rich and deep analysis of what works and what does not work in a care process. A SEIPS analysis provides input for designing and implementing a work system intervention.

TABLE 4.1
Examples of Questions on the Work System

UK Chartered Institute of Ergonomics & Human Factors (2021)	Adapted from Carayon et al. (2003, 2007a)
• *People*: who influences it?	• What are the characteristics of the *individual* performing the work? Does the *individual* have the physical, sensory and cognitive abilities to do the work? If not, can any of these be accommodated by other work system elements?
• *Task*: what is being done?	• What *tasks* are being performed? What are characteristics of the *tasks* that may contribute to health care safety? What in the nature of the *tasks* allows the individual to perform them safely or unsafely?
• *Tools and technology*: what is being used?	• What *tools and technologies* are used to perform the work? Do they increase or decrease the likelihood of adverse events?
• *Environment*: where is it being done?	• What in the *physical environment* promotes or hinders safety?
• *Organization*: how is it being completed?	• What in the *organization* prevents or allows exposure to hazard? What in the *organization* promotes or hinders safety?
• *External environment*: what outside of the organization may be of influence?	• What in the *external environment* promotes or hinders safety?

4.4.1 SEIPS-based Data Collection

Data collection on the work system typically involves multiple mixed methods, including observation, interview, focus group and survey. The data collection systematically addresses all work system elements to ensure that all aspects of the work system are considered and analyzed. Examples of questions on the work system elements are displayed in Table 4.1. The UK National Health Service (NHS) has developed a quick reference guide for SEIPS that includes a template and explorer questions for each of the work system elements; this can be found at: https://www.england.nhs.uk/wp-content/uploads/2022/08/B1465-SEIPS-quick-reference-and-work-system-explorer-v1-FINAL.pdf

4.4.1.1 SEIPS-based Observation

SEIPS-based observation often uses a data collection sheet organized with columns that represent different work system elements. Figure 4.2 shows the observation sheet used in our outpatient surgery study to shadow patients (Schultz et al., 2007). A similar data collection sheet can be used to observe any other work system and care process. Sometimes we add a column to keep track of time for specific events or activities. The HFE analyst writes notes in the form of text to describe the work system elements in each column of the observation sheet. The observation sheet can be paper or electronic.

Outpatient Surgery Patient Shadowing Form

Beginning time: Ending time:

INDIVIDUALS	TASKS	TOOLS & TECHNOLOGY	ENVIRONMENT	ORGANIZATION

FIGURE 4.2 Example of SEIPS-based Observation Sheet (Schultz et al., 2007).

Observations can be complemented by interviews as the observer asks clarifying questions to the person being observed; this is called contextual inquiry or mixing observation with clarifying questions on the spot (Beyer & Holtzblatt, 1999). The contextual inquiry method was, for instance, used to identify patient safety hazards in the perioperative period of cardiovascular surgery (Gurses, Kim, et al., 2012). The hazards were categorized in the work system elements; e.g. hazards in the *task* element were avoidable time pressure and unexpected changes, ambiguities due to different preferences of health care professionals and non-value adding tasks. Walk-Through-Talk-Through (WT3) is another approach that combines observation and interview. Brownhill and Bowie (2024) showed how SEIPS can be integrated in the WT3 approach and how the WT3 approach has been used to address health care safety, such as analysis of a blood sampling incident and analysis of medication administration by hospital nurses.

4.4.1.2 SEIPS-based Interview

SEIPS-based data collection via interview can be organized using an interview guide that asks the interviewee about their work system with follow-up questions on what works and what does not work in their work system. For example, we conducted interviews with care managers involved in the coordination of care for patients with chronic diseases (Carayon, Alyousef, et al., 2012). The interview began by asking care managers to describe a typical work day; this is a 'warm-up' step. Then the interviewer asked questions on work system barriers; see Table 4.2.

TABLE 4.2
Example of SEIPS-based Interview Questions in a Study of Care Managers (Carayon, Alyousef, et al., 2012)

Questions on work system barriers:

Please think of instances when you feel your performance was challenged or below par due to problems in your 'work system'. These instances may have been related to [present the list on a card]. For example, when you perform medication reconciliation, what can occur (or not occur) that can interfere with your ability to efficiently and accurately perform medication reconciliation?

- ☐ *Workspace or work environment.*
- ☐ *Resources (e.g. equipment, internet access, other workers).*
- ☐ *Information.*
- ☐ *Interactions with patients/family.*
- ☐ *Work schedule.*
- ☐ *Training.*
- ☐ *Policies and procedures.*
- ☐ *Goals and expectations.*
- ☐ *Other* ______________

Please briefly describe any such instance(s) you experienced by explaining the situation and what you think caused it. You may refer to the list above or identify other causes not listed.

Questions on work system facilitators:

Please think of instances when you were able to perform your job very well. These instances may have been related to [present the list on a card]. For example, when you perform medication reconciliation, what can occur (or not occur) that can interfere with your ability to efficiently and accurately perform medication reconciliation?

- ☐ *Workspace or work environment.*
- ☐ *Resources (e.g. equipment, internet access, other workers).*
- ☐ *Information.*
- ☐ *Interactions with patients/family.*
- ☐ *Work schedule.*
- ☐ *Training.*
- ☐ *Policies and procedures.*
- ☐ *Goals and expectations.*
- ☐ *Other* ______________

Please briefly describe any such instance(s) you experienced by explaining the situation and what you think caused it. You may refer to the list above or identify other causes not listed.

Care managers were then asked about work system facilitators, i.e. instances when they were able to perform their job very well. The same list of categories (e.g. workspace or work environment, resources, etc.) was used to probe about barriers and facilitators; these categories represent different work system elements.

4.4.1.3 SEIPS-based Focus Group

When collecting data on work systems and care processes, we can use focus groups and take advantage of the social interaction during group meetings to collect a broad and diverse range of data on work system barriers and facilitators. In focus groups with participants physically present, we typically display an image of the SEIPS model to anchor the conversation. For instance, we used focus groups to gather data on work system barriers and facilitators related to a CDI bundle: one focus group of hospital nurses (Ngam et al., 2017); and three separate focus groups of physicians, nurses and environmental services staff in an academic hospital (Musuuza et al., 2019). We also used a series of two focus groups with ICU nurses to analyze the safety of their medication management process (Faye et al., 2010). In the first focus group of seven ICU nurses, we identified a range of failure modes in the seven stages of the medication management process (i.e. things that could go wrong in the process): assessing the patient, obtaining a medication, administering a medication, monitoring/re-evaluating a patient, educating the patient and/or family during the ICU stay, educating the patient and/or family in preparation for discharge and conducting the nurse-to-nurse handoff. In the second focus group, that occurred two months later with five of the seven ICU nurses who participated in the first focus group, we used the work system model to help nurses identify the range of factors contributing to the failure modes for the various stages of medication management. This two-step proactive risk assessment (PRA) was informed by the SEIPS model and produced a deep and broad analysis of safety in the ICU nursing medication management process (Faye et al., 2010).

4.4.1.4 SEIPS-based Survey

In contrast to interviews and focus groups with a limited number of participants, a survey can reach many people and provide qualitative and/or quantitative data on work system barriers and facilitators. In our early outpatient surgery study, we used a survey with two open questions on work system barriers and facilitators (Carayon et al., 2005). The question on performance obstacles (or work system barriers) asked staff to identify 'instances when your performance was challenged or below par' and the question on facilitators asked staff to describe 'instances when you were able to perform your job very well'. We provided examples and categories of factors to consider; these represented all work system elements:

> your workspace or environment; resources needed to accomplish your job, such as equipment, other workers, or time available; information needed to do your work; your work schedule; the training provided to you; policies and procedures related to your work; goals and expectations given to you; or some other reason.

These open questions are similar to the interview questions shown in Table 4.2. The main difference between the interview and the survey is the inability to ask follow-up and clarifying questions with a survey.

A SEIPS-based survey contains open or closed questions. For an example of a survey with closed questions, see below the description of our study of work system barriers and facilitators experienced by ICU nurses (Gurses & Carayon, 2009; Gurses & Carayon, 2007).

4.4.1.5 Combining Data Collection

In SEIPS analyses, data collection methods are often combined. As described above, observations can be enhanced with on-the-spot interviews with the contextual inquiry method (Beyer & Holtzblatt, 1999). Observation and interview can be conducted separately, and the observation and interview data are then combined at the analysis stage. For instance, in a study of physicians, nurses and pharmacists in the ED of an Ethiopian hospital, Abebe and Chui (2018) first observed eight individuals for a total of 40 hours and then interviewed the same eight individuals. The combined analysis of observation and interview data identified multiple work system barriers to medication safety, such as limited supervision and support to junior physicians in the ED and physical separation of pharmacists from the rest of the ED staff that limited team communication.

Another example of combining methods is the use of a video of a care process followed by a SEIPS-based interview. In a study of family-centered rounds in a pediatric hospital, we video-recorded bedside rounds and then interviewed five health care team members and five parents who reviewed video records of their rounds to identify work system barriers and facilitators (Carayon, Li, et al., 2014). This stimulated recall method is similar to the collective confrontation method developed by French ergonomists (Mollo & Falzon, 2004). See more information on this project in Chapter 5 (case study #1).

4.4.1.6 Mixing Qualitative and Quantitative Data

SEIPS-based data collection can produce mixed data, i.e. qualitative (e.g. text) and quantitative (e.g. score on a questionnaire scale) data. Mixed qualitative and quantitative data can be gathered with separate data collection methods; e.g. an interview for qualitative data and a survey for quantitative data. Mixing of qualitative and quantitative data can also be done at the analysis stage; e.g. quantifying qualitative data such as interview data, which are analyzed to produce frequencies of reporting various work system barriers and facilitators (Carayon, Kianfar, et al., 2015).

As an example, I describe an early project, where we examined workload experienced by ICU nurses, which is known to affect health care safety (Carayon & Gurses, 2005; Tarnow-Mordi et al., 2000). We conducted an initial interview study of 15 nurses in a medical-surgical ICU of an academic hospital [qualitative data], which was followed by a survey of 272 nurses in 17 ICUs of seven hospitals [quantitative data] (Gurses & Carayon, 2009; Gurses & Carayon, 2007). In the initial study, we used a semi-structured interview guide and identified many barriers in the ICU nursing work system that contributed to workload and had a negative impact on ICU patient safety and nurses' quality of working life, such as insufficient work space, noisy environment and time spent looking for supplies (Carayon et al., 2005; Gurses & Carayon, 2009). These qualitative interview data were used to create a questionnaire about work system barriers that was administered in the follow-up survey study. The barriers reported by at least 30% of the surveyed ICU nurses included: noisy work environment (46%), distractions from families (42%), hectic (40%) and crowded work environments (37%), delay in getting medications from pharmacy (36%), spending considerable amount of time teaching families (34%), equipment not being available–someone else using it (32%) and patient rooms not well-stocked (32%) (Gurses & Carayon, 2007). A follow-up statistical analysis showed that the following work system barriers increased ICU nursing workload and affected their perceptions of safety and quality of working life: poor physical work environment, ineffective supply chain management (delay in getting medications from pharmacy, poorly stocked patient rooms, and disorganized supplies area), suboptimal access to patient charts and management of multiple family-related issues and patient admissions (Gurses et al., 2009). This study shows how different data collection methods can be used to analyze a work system and qualify and quantify barriers in the work system.

4.4.2 Qualitative Analysis of Work System Barriers and Facilitators

SEIPS analysis produces information on work system barriers and facilitators, which can influence health care safety. Data on work system barriers and facilitators often consist of qualitative data, i.e. text in the form of observation notes, transcription or notes of interview or focus group conversation or text responses to a survey. The qualitative data analysis of work system barriers and facilitators can produce a list of barriers and facilitators organized according to the work system elements. For each work system element, sub-themes or dimensions are identified. For instance, in a study of infection control in a large hospital in India, Barker, Brown et al. (2017) analyzed interview data collected from 20 nurses and physicians. For each work system element, they identified specific barriers and facilitators to effective infection control. For instance, for the *tools and technology element*, discomfort of masks and lack of comprehensive EHR were barriers, whereas ample infection control supplies and an electronic database for laboratory results were facilitators. Many studies and projects map barriers and facilitators to the work system elements; however, this is not always the case.

The qualitative data analysis of work system barriers and facilitators can lead to emergent dimensions that represent interactions between the work system elements. Often, an inductive approach driven by the work system model is combined with a deductive approach, e.g. a thematic analysis. In a study in three pediatric units in the United Kingdom, Sutherland et al. (2024) use observation and interview with nurses, physicians, pharmacists and parents to characterize the work system factors that influence medication safety. They first conducted an inductive data analysis that produced three themes: physical space, situational space and cognitive space, which were deductively connected to the work system elements. A second inductive data analysis identified specific factors within each of the three themes. For instance, the theme of 'physical space' was connected to the work system elements of *tools & technology, environment* and *task*. Specific inductively driven categories of the theme included: workstations and spaces, storage and access and distractions and interruptions.

When analyzing work system barriers and facilitators, sometimes the emergent dimensions or themes are both positive (i.e. facilitator) and negative (i.e. barrier); but this is not always the case. Some dimensions may either relate to barriers or to facilitators. For instance, in a mixed-methods study of the impact of secure messaging on the workflow in five primary care clinics, we found eight dimensions related to the work system (e.g. amount of work, usability and organization of work) and three dimensions related to the outcomes (i.e. satisfaction with technology, impact of technology on patient satisfaction and impact of technology on quality and safety of care) (Hoonakker et al., 2017). The interviewees talked about the usability of secure messaging technology as either a barrier (e.g. a nurse experiencing difficulty in managing multiple messages from a patient) or as a facilitator (e.g. a patient describing the ease of use of the secure messaging technology that is embedded in the patient portal). The dimension of 'inappropriate use' was, of course, only a barrier.

An analysis of work system barriers and facilitators using various data collection methods based on SEIPS produces input for designing interventions. A thematic analysis of work system barriers and facilitators can contribute to the definition of user requirements, such as how to design community-based diagnostic services (Sujan, Crumpton, et al., 2024). Section 4.5 provides additional information on the use of SEIPS for designing interventions.

4.4.3 Examples of SEIPS Analysis of Jobs and Roles

SEIPS has been largely used to analyze specific jobs or roles, such as ICU nurses (Gurses & Carayon, 2009; Gurses & Carayon, 2007), tele-ICU nurses (Hoonakker et al., 2013; Hoonakker & Carayon, 2018) and community pharmacists (Chui et al., 2012). The focus of SEIPS on specific jobs has also helped to understand interactions between the external environment and the work system, such as the impact of the COVID pandemic on nursing jobs (Bethel et al., 2022). SEIPS can also be used to analyze specific job tasks and roles, such as documentation by ED physicians (Fouquet et al., 2020) and endotracheal tube cuff pressure management by ICU nurses (Kebapci & Ozkaynak, 2022).

While SEIPS has been extensively used to analyze a range of health care jobs, it has also been applied to understand the 'work' of patients with various conditions and in multiple settings. For instance, McLean et al. (2024) used the SEIPS model to describe factors that affect access to health care for autistic adults. Using interviews of 23 autistic adults in the United States, they characterize work system factors that either hinder or facilitate access to health care for autistic adults. A *technology*-related facilitator was the ability of the autistic adult to use an online system to schedule an appointment. The *physical environment* in health care settings was critical for autistic adults' decision to visit a doctor; fluorescent lighting, smells and noise could create discomfort and bother them. Another example of the application of SEIPS to patients is a study of people living with cystic fibrosis having lifelong medication regimens that, if not well-designed and monitored, can lead to complications, such as infections (Abraham et al., 2021). For instance, people living with cystic fibrosis identified their interaction with pharmacists (i.e. characteristic of the *organization element*) as key to helping them manage their complex medications.

SEIPS has been increasingly used to analyze formal and informal jobs and roles in health care, including those of patients and care partners (Carayon, Albayrak, et al., 2021). An important effort connected to SEIPS is the development of models of patient work in the HFE discipline (Holden & Valdez, 2021; Valdez & Holden, 2021). The SEIPS family has 'cousins' in the form of patient work models (Holden, Schubert, & Mickelson, 2015; Holden et al., 2017; Valdez et al., 2015). The Patient Work System model includes the following elements: *patient, tasks, tools and technologies, physical environment, social environment and organizational environment*

(Holden, Schubert, & Mickelson, 2015; Valdez et al., 2015). It has been used to improve the design of technologies and their integration in patients' lives (Valdez et al., 2015). Holden, Schubert et al. (2015) used the Patient Work System model to identify barriers reported by patients with acute heart failure in their self-care, such as complexity and time demands of self-care (i.e. *task element*) and transportation and insurance problems (i.e. *organization element*).

SEIPS and the Patient Work System model have also been used to analyze the work of care partners, such as care partners of elderly patients with chronic heart failure (Holden, Schubert, & Mickelson, 2015) and informal care partners who help older adults with medication management (Look & Stone, 2018). See Werner, Ponnala et al. (2020) for a review of applications of SEIPS and the Patient Work System model for understanding and improving the work of patients and care partners. The extension of SEIPS and the SEIPS family to understand and improve the work of patients and care partners is important.

4.4.4 Examples of SEIPS Analysis of Care Processes

SEIPS integrates the work system model of Smith and Carayon (Carayon, 2009; Smith & Carayon-Sainfort, 1989); but SEIPS is more than the work system model. SEIPS shows how care processes are clearly connected to the work system. A care process is not just a series of tasks; it is a combination of the work system elements that involves multiple individuals, roles, units and sometimes organizations. We have applied SEIPS to analyze various care processes, e.g patient rounds in a pediatric hospital (Carayon, Li, et al., 2014; Kelly et al., 2013) and care transitions between ED, OR and Pediatric Intensive Care Unit (PICU) (Wooldridge et al., 2020). See Chapter 5 for additional information on these projects.

SEIPS has been used to analyze a range of processes, such as the decontamination and assembly stages of instrument sterile processing (Alfred et al., 2020, 2021), the MRI work process (Pickup et al., 2019) and safety in the care process of newborns (English et al., 2021). Kirkendall et al. (2023) developed SEIPS-based process maps to identify vulnerabilities and potential for adverse events in medication management at home for children with type 1 diabetes. SEIPS analysis of processes can identify work system barriers and facilitators and patient safety risks at different stages of a process, for instance, in the outpatient surgery process (Carayon, Hundt, Alvarado, et al., 2006), in the anesthesia medication use process (Coppola et al., 2022) and in the medication administration process in the pre-hospital setting (Odberg et al., 2025).

The HFE discipline and the QI field have developed numerous process analysis methods. Some of these methods are easy to use by health care staff. See, for example, a study comparing eight process analysis methods that provide different visualizations of processes (Jun et al., 2009); flowcharting was considered a highly usable method. Process (workflow) analysis methods are critical in the context of health IT implementation (Carayon, Cartmill, et al., 2012). Some process analysis methods specifically integrate the work system model and its elements, such as the SEIPS-based process modeling method (Wooldridge et al., 2017) and the SEIPS-based patient journey mapping method (Wust et al., 2022). The SEIPS-based process modeling method is described in Section 3.3.2 of Chapter 3. It has been used to analyze processes within a primary care facility (Wooldridge et al., 2017), the medication administration process in a nursing home (Odberg et al., 2020), the care transition between hospital and home care for sepsis survivors (Oh et al., 2024) and the processes for detecting and managing patient deterioration in a virtual hospital (Bamgboje-Ayodele et al., 2025).

4.4.5 SEIPS-based Patient Journey Mapping

A patient journey depicts the care process from the perspective of the patient; this is a core concept in the SEIPS 3.0 model (Carayon, Wooldridge, et al., 2020). Patient journey mapping has been recommended as a critical method to use in the initial stage of HCD (Melles et al., 2021). It helps to analyze the different steps of patient work, such as the work involved when a patient is having hand/wrist surgery (De Ridder et al., 2018) or when a patient is having a diagnostic gastrointestinal test with video technology (Simonse et al., 2019).

We developed the SEIPS-based patient journey mapping to ensure that the patient journey process map was systematically analyzed and included all work system elements (Wust et al., 2022). The work system elements are integrated in a SEIPS-based patient journey map as follows:

- *Person*: the map is created from the perspective of the patient and includes all other people involved, such as a nurse or care partner.
- *Tasks* are represented throughout the patient journey map, although not necessarily in a sequential and linear manner.
- *Tools and technology* are represented together with the tasks, sometimes with icons.
- *Organization* is represented as the flow and the structure of the journey. A journey map may also include information on the organizational context that applies to the entire map; this is often included as a 'note' to the map.
- *Environment* is represented as different steps of the map as the patient interacts with different physical environments in various health care units, organizations or settings.

A SEIPS-based patient journey map does not necessarily show activities in a linear manner. In line with the concept of complex adaptive systems (see Chapter 2), we know that care processes involve adaptation, feedback, iteration, etc.; therefore, a patient journey map is rarely a sequential and linear process. There are sometimes sequential activities as one activity cannot happen before another one, but many activities are not temporally or linearly organized. SEIPS-based journey mapping has been used to describe the experience of an adult in the ED (Wust et al., 2022) and the experience of children with mental and behavioral conditions in the ED (Soman et al., 2025).

4.4.6 Reflection on Analyzing Work – Work-As-Done (WAD) Versus Work-As-Imagined (WAI)

The concepts of WAD versus WAI are extensively discussed in the safety literature (Hollnagel et al., 2006). The concepts of WAD and WAI have been proposed by researchers in the resilience engineering domain (Braithwaite et al., 2015; Clay-Williams et al., 2015; Hollnagel, 2014). Analyzing work (the actual work performed or so-called WAD) is what HFE professionals do and have done for many years. There are numerous HFE books on task analysis (Kirwan & Ainsworth, 1992), evaluation of work (Wilson & Corlett, 2005), activity analysis (Guerin et al., 2006), etc. The deep HFE knowledge of work analysis and its various methods need to be integrated when discussing WAD and WAI. New methods, such as FRAM, have emerged from the literature on resilience engineering (Hollnagel, 2012b); they complement and add to the toolbox of HFE methods aimed at analyzing and understanding the actual work.

The conceptualization of WAD and WAI could benefit from the conceptual work of French ergonomists who early on distinguished 'tâche' (task) and 'activité' (activity) and prescribed and real work (Daniellou, 2005; Leplat, 1989). This distinction between prescribed and real work has led to research and scholarly development on the analysis of work activity and the creation of HFE methods for activity analysis, such as 'confrontation' methods (Mollo & Falzon, 2004) with applications in health care (Carayon, Li, et al., 2014; Mollo & Falzon, 2008; Nascimento & Falzon, 2012). The French ergonomics research tradition also focuses on understanding strategies used by workers to do their (real) work while dealing with constraints and demands in their work environment (often imposed by prescriptions and rules of the organization, or the workers' interpretation of those rules) (Faye & Falzon, 2009). It is important to understand how workers think about their activities, the decisions they make about what to do (and what not to do), their interpretation of what is asked of them by the organization, etc. The terms WAD and WAI are increasingly used, but we should not overlook the research in HFE, including the French ergonomics tradition and its concepts of 'tâche' (task) and 'activité' (activity). Health care safety researchers and practitioners should become familiar with the rich HFE literature, including the French ergonomics literature (Barcellini et al., 2024; Falzon, 2015; Guerin et al., 2006).

4.5 SEIPS FOR DESIGNING AND IMPLEMENTING AN INTERVENTION

SEIPS has been extensively used in analyzing health care work systems and processes. The results of the SEIPS analysis can provide input to decide what to change in the work system, i.e. the intervention. SEIPS should also be considered when designing and implementing an intervention. An intervention always has an impact on the entire work system, its elements and their interactions; therefore, SEIPS can be used to describe an intervention. Accordingly, HFE knowledge should be incorporated in the design and implementation of systems change in order to improve health care safety (Wood & Wiegmann, 2020).

BOX – SEIPS AND DESIGN OF WORK SYSTEM INTERVENTIONS

A SEIPS analysis provides information on what needs to be changed in the work system with the goal of improving health care safety. As a SEIPS analysis identifies what works and what does not work, this information is critical to decide what and how to redesign the work system and the care process. In addition, during data collection for a SEIPS analysis, information can be collected on ideas for work system redesign. SEIPS can also be used to map potential solutions to the work system model and its elements. An intervention aimed at improving health care safety always involves changes in the entire work system; therefore, SEIPS can be used to describe the intervention and its impact on the entire work system.

4.5.1 SEIPS Analysis as Input to Design of Intervention

A SEIPS analysis produces information that can serve as input to the design of interventions aimed at improving health care safety. SEIPS analysis of work system barriers and facilitators produces information on what needs to be changed in the work system. We should do everything we can to support the work of health care staff, patients and care partners. Therefore, work system barriers should be eliminated or mitigated; facilitators should be kept or enhanced (see the discussion on the Balance Theory in Chapter 3). For instance, in a study of 22 hospital nurses (Steege & Dykstra, 2016), nurses identified factors contributing to their fatigue as different dimensions in each of the work system elements. Within the *physical work environment*, three themes were identified: unit layout (e.g. distance to location of supplies), break room layout and patient room layout. In addition, nurses identified solutions for reducing fatigue in each of the work system elements. In the *physical environment*, nurses talked about improving the design of the break room to allow for quiet relaxation and rest.

Another example of the use of SEIPS-based analysis for the design of safety interventions is provided by Sujan, Crumpton, et al. (2024). Whereas some studies show how interventions are linked to specific work system elements (Schneider et al., 2019; Steege & Dykstra, 2016), the study of Sujan, Crumpton, et al. (2024) goes beyond the analysis of specific work system elements and associated interventions; they identify themes (i.e. combination of work system elements) that represent opportunities and considerations for designing interventions. In a project aimed at informing the design of community-based cardiac diagnostic services, they used multiple methods to collect data, including observation, interviews and workshops. Different roles were involved in the data collection: e.g. physiologists, cardiologists and administrators. An initial analysis identified barriers and facilitators within each of the work system elements, which was followed by a thematic analysis of interactions between work system elements. This led to the identification of three design themes: (1) to promote autonomy and professional growth (e.g. promoting autonomy of physiologists), (2) to focus on needs of patients in the community (e.g. moving services closer to the patient) and (3) to facilitate communication across organizations (e.g. improving digital sharing of patient-related information). These design themes provide important information on user requirements for the efficient, effective and safe design of community-based cardiac diagnostic services.

Data collection for SEIPS analysis may also include data on potential solutions. When collecting data for a SEIPS analysis, the participants who provide information on work system barriers and facilitators can also share their ideas for improving the work system and care process. Let us look at one example of safety for complex surgical patients. Patients who undergo complex surgeries are sometimes re-admitted to the hospital because of the poorly designed process of transition of care from hospital to home. How can we improve that transition to avoid unnecessary re-hospitalizations? Acher et al. (2015) interviewed 12 patients who had complex abdominal surgery of the pancreas, liver, colon, rectum or esophagus and were re-admitted to the hospital within 30 days of discharge. The researchers also conducted a focus group with surgical residents, nurses, a care coordinator and a pharmacist. The qualitative interview and focus group data were analyzed to identify the reasons for the poor preparation of patients for discharge and their hospital re-admission; the reasons were mapped to the work system elements. For each work system element, one or several themes were identified, such as 'Aspects of the content of educational materials to be used at home that impeded teaching, understanding, or use' (*technology element*). Finally, for each theme, several interventions were identified. Regarding the educational materials given to patients, interventions included presenting only patient-specific information in the materials and conducting practice sessions with patients. This example shows how SEIPS can be used for analyzing the work system (i.e. input to design the intervention) as well as for designing a set of interventions.

SEIPS has been used extensively to address HAIs around the world; see Chapter 1. Hand hygiene is an important component of interventions aimed at reducing HAIs; it can be even more effective when SEIPS is used to identify barriers to hand hygiene in specific care settings. This was demonstrated in an academic hospital in Ethiopia after implementation of the WHO guidelines for hand hygiene (Berman et al., 2021). Following a SEIPS analysis of barriers to hand hygiene, additional interventions were implemented to enhance the effectiveness of the WHO guidelines and further reduce HAIs. The SEIPS analysis was based on interviews with 22 health care professionals (i.e. physicians, nurses and midwives) in various hospital units and services. The qualitative interview data analysis produced information on barriers and facilitators in each of the work system elements. In the *tools and technology element*, barriers included a lack of functional water facilities and water supply, hand rub and soap, and gloves; facilitators included pocket hand rub, dispensers providing access to hand rub and a high number of sinks in hospital wards. The SEIPS-based data collection provided information on additional areas of improvement to enhance hand hygiene and the implementation of the WHO guidelines. A SEIPS analysis can help to improve the implementation of interventions, such as guidelines.

4.5.2 SEIPS for Describing the Intervention and Its Impact on the Work System

After completing a SEIPS analysis, various ideas for interventions arise. Different methods can be used to prioritize the intervention ideas, such as PRA (Proactive Risk Assessment) (Faye et al., 2010; Hoonakker, Carayon, et al., 2019), the Delphi technique (Rutkowski et al., 2025) or a survey on impact and feasibility (Xie, Carayon, Cox, et al., 2015; Xie et al., 2012). An intervention represents a change in an element or a part of the work system or the care process: e.g. a new technology or a new process. An intervention is never 'just … [fill in the blank]'; it is always a change in the entire system. Therefore, it is important to pay attention to the design of the intervention itself, as well as its fit (or lack of fit) in the work system and the care process or workflow.

SEIPS has been used to describe interventions aimed at improving health care safety, such as an intervention aimed at reducing the use of broad-spectrum antibiotics in the fluoroquinolone (FQ) class for ICU patients in order to prevent and reduce C. difficile infections (CDI) (Carayon, Thuemling, et al., 2021). The intervention consisted of *people* (i.e. junior and senior ICU physicians, infectious disease physicians and pharmacists) collaborating on multiple *tasks* (e.g. ICU physician prescribing antibiotics and talking with an infectious disease physician for advice on alternative medications). The prescribing physician uses various *tools and technology*, such as the EHR. Prescribing work is performed in the ICU (*physical environment*) and during rounds (*organization*). This example shows how the work system model and its elements can be used to describe an intervention; this helps to ensure that the intervention is systematically and fully described. A similar use of SEIPS was done in a review of interventions aimed at improving quality and safety of care in long-term care facilities (Peryer et al., 2022). The interventions described in the 33 studies of this systematic review were mapped to the work system elements; this analysis showed that most barriers addressed by the studies were related to the *task and organization elements*.

Another way of using SEIPS to describe interventions is to map them to barriers in the work system. In a study aimed at improving working conditions and mental well-being for ED physicians as well as the quality of ED care, Schneider et al. (2019) identified barriers in each of the work system elements; these barriers were then addressed with specific solutions. For the *organization* barrier of a lack of breaks during work time, some of the solutions included scheduling additional nurses and short-term rotation across units for short-term replacement of nurses taking a break. In a similar manner, the work system model was used to describe various interventions for each element of the work system with the aim of improving the safety of inter-hospital transfers (Yu et al., 2024). For instance, in the *organization element* of the work system, interventions for safe inter-hospital transfers included: improving shared situation awareness among the sending and receiving teams and using call centers to support information flow and care coordination.

The SEIPS 2.0 model was adapted to develop interventions for increasing in-hospital mobility (Zisberg et al., 2018). The adapted model helped to understand all work system elements involved in the process of in-hospital mobility and defined outcomes for patients (e.g. 900 steps a day and maintaining functional status) and for health care staff and the organization (e.g. satisfaction with care). The adapted model was used in a series of workshops and focus groups with health care staff and organizational leaders to create an action plan that addressed work system barriers to in-hospital mobility.

Other examples of the use of SEIPS for describing interventions and their impact on the work system and outcomes can be found in the case studies described in Chapter 5.

4.5.3 SEIPS and Integration of HFE Design Principles in Designing an Intervention

When designing an intervention, knowledge from HFE should be integrated. HFE knowledge is vast as demonstrated by the size (1,600 pages) of the most recent edition of the 'Handbook of Human Factors and Ergonomics' (Salvendy & Karwoski, 2021). Here, I provide some examples; for additional examples, see Chapter 3 and the case studies in Chapter 5. HFE knowledge often comes in the form of design principles or criteria, which can be categorized as physical, cognitive or organizational HFE.

4.5.3.1 Examples of Physical HFE Design Principles

In the area of physical HFE, principles are available for the design of workspaces (e.g. patient room, medication room). Specific physical HFE guidelines exist for the workspaces of surgeons during open, laparoscopic and robot-assisted surgery; this physical HFE knowledge aims to ensure good posture of surgeons and avoid WRMDs (e.g. injuries of the neck, back or shoulders) (Tetteh et al., 2024). Hignett and Wolf (2016) provide physical HFE guidance on how to prevent and reduce patient falls in the hospital. Reducing the physical load of health care work, in particular nursing work, is key to worker safety. Principles include avoiding frequent bending and twisting and heavy physical work when nurses care for patients and move them (Waters, 2010). Providing well-designed lifting aid devices and other HFE-based interventions for patient handling can reduce health care worker injuries (Fujishiro et al., 2005; Hignett et al., 2013). Allread and Israelski (2010) provide physical HFE design principles based on knowledge in anthropometry and biomechanics, which can be applied to the design of medical devices and equipment.

4.5.3.2 Examples of Cognitive HFE Design Principles

Health IT are interventions that can significantly benefit from cognitive HFE knowledge, such as the Nielsen–Shneiderman usability heuristics of Zhang et al. (2003), the heuristics of Scapin and Bastien (1997) or specific usability heuristics for patient-oriented documents (Barton et al., 2023; Tremoulet et al., 2018). In several projects, we have developed HFE design principles for multiple types of health IT, e.g. team health IT (Carayon et al., 2022). Cognitive HFE design principles for CDS technology include automation of information acquisition and analysis, support of decision selection, minimizing workload, consistency and visibility (Carayon, Hoonakker, et al., 2020).

4.5.3.3 Examples of Organizational HFE Design Principles

In the area of organizational HFE, multiple design principles exist for teamwork (Salas et al., 2008), team training for patient safety (Salas & Rosen, 2013) and work schedules and shiftwork (Rogers, 2012). For example, the 2012 IOM report describes five principles for teamwork: shared goals, clear goals, mutual trust, effective communication, and measurable processes and outcomes (Mitchell et al., 2012). In Chapter 3, I provide information on principles for designing 'good' jobs from the STS Theory (Davis, 1977; Trist, 1981); see Table 3.5 in Chapter 3. The principles of High Reliability Organization (HRO) should also be incorporated in the toolbox of organizational HFE: sensitivity to operations, commitment to resilience, preoccupation with failure, reluctance to simplify and deference to expertise (Weick & Sutcliffe, 2001). The HRO principles can be applied at many system levels, including the team level (Wilson et al., 2005).

In this section, I provide only a sample of HFE design principles and examples of their application to health care safety. In a paper commissioned for the 2013 AHRQ Patient Safety Practices Report (Carayon et al., 2013; Carayon, Xie, & Kianfar, 2014), we describe multiple examples for the integration of HFE in health care safety interventions, such as the design of radiotherapy treatment and the design of an ED telemetry system. For further information on HFE design principles, one should consult HFE handbooks and books, and other resources; several of these are mentioned in this book. Integrating this vast HFE knowledge in designing a work system change and interventions on the care process is critical and benefits from the involvement of qualified HFE professionals. See discussion on embedding HFE professionals in health care in Section 2.5 of Chapter 2.

4.5.4 SEIPS and Integration of Organizational Principles in Implementing an Intervention

An intervention introduces a change in the work system and that change process needs to be managed. Differences in how an intervention is implemented in various health care organizations or units can explain its impact and integration in the work systems and care processes (Carayon, Thuemling, et al., 2021; Hay, Klonek, & Parker, 2020). The literature on change management and organizational change is extensive; we know a lot about 'good' implementation principles. For instance, Korunka et al. (1993) identified three key principles of good implementation of manufacturing technologies: project organization, participation of employees and training. Karsh (2004) conducted an extensive review of literature and developed a set of eight implementation principles for health IT: top management commitment, responsibility and accountability, structured program, training, pilot testing, communication and feedback, simulation and end-user participation. We consolidated the implementation principles of Korunka et al. (1993) and Karsh (2004) into five categories (Carayon, Alyousef, & Xie, 2012):

1. top management commitment,
2. stakeholder participation,

3. communication and feedback,
4. learning and training and
5. project management.

These five principles can be used to ensure that a work system change is well implemented and accepted. For each implementation principle, several activities can be identified and planned. These principles guided the successful implementation of family-centered rounds in a pediatric hospital (Xie, Carayon, Cox, et al., 2015). We organized several activities for each principle. For instance, the principle of top management commitment was instantiated as: asking the hospital medical director to participate in the design/implementation team and providing regular updates to a steering committee composed of several organizational leaders. For more information on this project, see case study #1 in Chapter 5. The implementation principles have also been used in the design and implementation of a patient-centered ED discharge process for older adults (Carayon, Wust, et al., 2025). For more information on this project, see case study #3 in Chapter 5.

4.6 SEIPS FOR EVALUATION OF AN INTERVENTION

SEIPS can be used to evaluate an intervention. Information from the evaluation can then provide input to improve the intervention in a continuous design process; this represents the feedback loops in SEIPS. A SEIPS-based evaluation identifies pluses and minuses of the intervention, i.e. work system barriers and facilitators related to the intervention and its integration (or lack of integration) into the care process. The evaluation can also include information on the impact of the intervention on outcomes, i.e. impact on patients (e.g. patient safety, patient experience) and on health care workers (e.g. workload, frustration or satisfaction, quality of working life). In this section, I provide examples of how SEIPS has been used to evaluate technologies and process interventions in various domains of health care safety: medication safety, communication and coordination in care transitions, diagnostic safety and HAIs.

4.6.1 SEIPS Evaluation of Technology Implementation

Health IT is often implemented as an intervention to improve patient safety; however, results are mixed (Brenner et al., 2016), including for health IT aimed at improving medication safety (Gates et al., 2021). Usability of health IT is a major factor that influences whether the technology will lead to improvement or deterioration of patient safety (Ratwani et al., 2018). Poor health IT usability is also a predictor of health care professional stress and burnout (Melnick et al., 2019). Improving the usability of health IT is critical, but not sufficient to produce benefits in health care safety; we need to consider the entire work system in which the technology is implemented and how it is integrated in the workflow and care process (Beuscart-Zephir et al., 2012; Carayon & Hoonakker, 2019).

SEIPS and, in particular, the work system model have been used to evaluate health IT and its role in health care safety. A systematic review of 55 papers identified a range of factors in the work system that influence safety associated with health IT

(Salahuddin & Ismail, 2015). Thirty-seven of the 55 studies mentioned factors in the *organization element* of the work system, including teamwork, training and organizational resources.

In a study of the implementation of EHR technology in two ICUs of a large hospital (Carayon, Wetterneck, et al., 2021), we found an overall reduction in harm associated with medication errors; however, while several medication errors were reduced (e.g. medications not dispensed or dispensed late) or eliminated (e.g. illegible orders), other medication errors increased, in particular duplicate medication ordering errors. We conducted a follow-up SEIPS analysis of the duplicate medication ordering errors that occurred after EHR implementation and found issues in all of the work system elements and their interactions (Wetterneck et al., 2011). Our study confirmed that lack of usability was a contributing factor, in particular, the poor design of alerts indicating duplicate medication orders (*technology element of the work system*). An interesting contributing factor was role distribution within the ICU team, i.e. the *organization element* of the work system. During ICU bedside rounds, the same order could be placed by two individuals, e.g. the attending physician (senior physician) and the medical resident (junior physician), within 1 to 10 minutes. This example shows how the *technology element* of the work system interacted with the *organization element* to create a hazardous situation that led to duplicate medication orders and potential medication-related harm to vulnerable ICU patients. We shared the SEIPS analysis of duplicate medication orders with the hospital and the EHR implementation team, which led to multiple changes, e.g. improving the usability of alerts and clarifying team roles during ICU rounds and the responsibility for order entry and supervision. This is an example of improvement and learning in the SEIPS feedback loops.

An in-depth analysis of the medication administration process in a hospital showed unanticipated effects of bar code medication administration (BCMA) technology that could potentially affect medication safety (Carayon, Wetterneck, et al., 2007d). Using an observation form organized around the work system elements, we collected data on 62 instances of medication administration where nurses used the BCMA technology. A surprising and unanticipated finding was the fact that, in 23 observations (37%), nurses completed the documentation of the medication administration before the medication was actually administered; this frequent work-around in the task sequence was potentially unsafe and was influenced by several work system factors, including time pressure and fragmented workflow. Systematically identifying work system barriers and facilitators related to the implementation and use of health IT is critical for its continuous improvement. Another example of SEIPS-based evaluation of health IT is provided in case study #2 in Chapter 5.

4.6.1.1 AI Technology in Health Care Work Systems

Artificial intelligence (AI) technologies are increasingly implemented in health care (National Academy of Medicine (NAM), 2025). We need to ensure that AI technologies are designed to fit the care process, do not add burden to busy workers and patients and do not create additional hazards in the clinical workflow. Beede et al. (2020) used multiple methods (observation and interview) to evaluate the implementation of an AI technology aimed at improving the diagnosis of diabetic retinopathy in several clinics in Thailand. Results show that multiple work system factors hindered the use of the AI technology. For instance, the poor lighting of the *physical environment* affected the quality of pictures taken of patients' eyes, which then limited the use of the deep learning algorithm embedded in the *technology*.

SEIPS can be applied to analyze the fit (or lack of fit) of AI technologies in work systems and care processes (Salwei & Carayon, 2022). In a study of a CDS aimed at supporting the diagnosis of PE in the ED, we identified both positive and negative interactions of the AI technology (i.e. the CDS) with the rest of the work system. For instance, the CDS supported the collaborative work between senior ED physicians and physicians-in-training (*interactions between people, tasks, technology and organization*). On the other hand, because the CDS was not available on a portable device, ED physicians could not use it while walking between patients and in the ED (*interactions between tasks, technology and physical environment*). In addition to analyzing the impact of AI technologies on work systems, we also need to pay attention to other HFE issues. As AI technologies integrate automation, HFE experts have called for attention to HFE issues of decision-making, trust and transparency (Endsley, 2023; Sujan et al., 2022).

4.6.2 SEIPS Evaluation of Process Interventions

Some interventions are focused on one particular element of the work system, such as health IT, while other interventions aim at changing care processes; this is what I call process interventions.

An Australian study assessed various outcomes of a program aimed at facilitating cross-specialty collaboration to improve the diagnosis and management of rare disease patients (Hay, Klonek, & Parker, 2020). The main features of this program were the introduction of regular expert panel meetings that allowed a range of specialists and clinicians to discuss currently undiagnosed patients and the creation of a new role, i.e. a nurse coordinator, who helped with information sharing and coordination of patients through the program. A SEIPS evaluation of the program relied on multiple data sources, i.e. documents, interviews and observations. One year after its implementation, the program resulted in a major increase in the percentage of families receiving a definitive diagnosis (55%). Using the concept of configuration of the SEIPS 2.0 model (Holden et al., 2013), the researchers compared the work systems before and after the implementation of the program. For instance, before the implementation of the program, regarding the *task element of the work system*, whereas task significance was high (i.e. providing diagnoses for rare disease patients), task identity (i.e. fragmented tasks across multiple specialties) and feedback (i.e. lack of feedback regarding one's clinical contribution to the final diagnosis) were low. After the implementation of the program, task significance was even higher and task identity and feedback increased significantly. The clinicians participating in the expert panel meetings shared successes in diagnosing patients (i.e. task significance), were able to participate in the entire process (i.e. task identity) and received frequent feedback on test results they had suggested (i.e. feedback). The visual comparison of the work system configurations showed greater integration and coordination of the various specialists and clinicians after implementation of the program. Some barriers remained after the implementation of the program, such as time pressures on clinicians from the *external environment* because of issues with health care funding, workforce and staffing. In line with the Balance Theory (Carayon, 2009; Smith & Carayon-Sainfort, 1989), the *organization and task elements* of the work system in the new program compensated for the negative influence of the *external environment*: the nurse coordinator significantly reduced the administrative work performed by clinicians and facilitated the integration and sharing of patient-related information. This process intervention led to positive outcomes for both patients and health care professionals.

In Section 4.5.2, I showed how the work system model was used to describe a process intervention aimed at improving antibiotic stewardship in hospitals (Carayon, Thuemling, et al., 2021); the intervention reduced prescriptions for antibiotics (i.e. fluoroquinolones or FQ) that can lead to CDI (Carayon, Thuemling, et al., 2021; Tischendorf et al., 2020). While the intervention had a positive safety outcome, a work system analysis showed that its design could be improved (Tischendorf et al., 2020). Interviews with physicians, advanced practice providers (APPs) and pharmacists showed both barriers and facilitators in several work system elements. For example, the *organizational culture and structure*, in particular the strong hospital's stewardship program, facilitated the implementation of the intervention. Another *organizational facilitator* was the 24/7 presence of pharmacists in the ICU. On the other hand, the poor visibility of the antibiogram in the EHR was a *technology barrier* to prescribing alternative antibiotic medications. The follow-up SEIPS-based analysis provided input to the next phase of improvement for the design and implementation of the intervention.

Medication safety is a major issue for older adults in community pharmacies, especially when over-the-counter (OTC) medications are used inappropriately. An interdisciplinary team designed, implemented and evaluated an intervention aimed at improving communication between pharmacists and older adults through a physical redesign: selected high-risk OTC medications were made more visible and moved closer to pharmacists, and signage and aisle redesign provided clearer safety warnings to older adults (Gilson et al., 2021). Interviews with eight pharmacists and five technicians in four pharmacies provided overall positive feedback on the process intervention, including its integration in the work system and outcomes of patient engagement in safe OTC medication use and efficient interactions with pharmacy staff. A negative aspect of the intervention was related to the *technology element*, i.e. the signage was not always noticed by the older adults, according to the pharmacy staff. Another example of SEIPS-based evaluation of process interventions is provided in case study #1 on family-centered rounds in Chapter 5.

Results of SEIPS-based evaluations provide important information that feeds back to the work system and the care process. As discussed earlier, a key feature of systems thinking is emergence; therefore, it is difficult (impossible?) to anticipate all changes and safety consequences when designing and implementing an intervention. The feedback loops use data from the SEIPS evaluation and help to continuously improve the work system, care process and outcomes.

4.7 SEIPS FOR EVENT ANALYSIS

When SEIPS is used for event analysis, typically the work system model is applied to identify factors that were involved in the adverse event. For instance, the work system model is used to investigate adverse events in Dutch hospitals (Baartmans et al., 2022); an analysis of 69 events reported by 28 hospitals shows how elements of the work system and their interactions are involved in adverse events (Baartmans et al., 2023).

In the United Kingdom, SEIPS has been used in several investigations by the Healthcare Safety Investigation Branch (HSIB) (Weaver et al., 2021) that transitioned to the Health Services Safety Investigation Body in 2023. Several HSIB reports show how SEIPS can be used to analyze patient safety events, such as delayed diagnosis in primary care (https://www.hssib.org.uk/patient-safety-investigations/continuity-of-care-delayed-diagnosis-in-gp-practices/) and assessment of VTE risks in pregnancy and during the postnatal period (https://www.hssib.org.uk/patient-safety-investigations/the-assessment-of-venous-thromboembolism-risks-associated-with-pregnancy-and-postnatal/). Conducting a systems analysis of adverse events is not easy; therefore, it is important to develop materials to help investigators apply systems approaches to analyze adverse events, such as SEIPS. HSIB has developed a SEIPS toolkit for investigators, which can be found at: https://www.hssib.org.uk/news-events-blog/the-investigators-toolkit-seips/.

In 2020, the UK NHS proposed the Patient Safety Incident Response Framework (PSIRF), which promotes learning from patient safety incidents using systems approaches. SEIPS is recommended by PSIRF. As part of the PSIRF toolkit, a reference guide on SEIPS was published in 2022 and is widely available at: https://www.england.nhs.uk/wp-content/uploads/2022/08/B1465-SEIPS-quick-reference-and-work-system-explorer-v1-FINAL-1.pdf. The reference guide includes a PowerPoint template of SEIPS that can be used in investigations or analyses.

It is important to note that many other HFE and systems frameworks and models have been specifically designed for adverse event analysis and used in health care, such as AcciMap, HFACS, STAMP and the London Protocol (Cohen et al., 2018; Isherwood & Waterson, 2021; Leveson et al., 2020; Taylor-Adams & Vincent, 2004; Vincent et al., 2025). An interesting suggestion is to combine frameworks and models for adverse event investigation, such as SEIPS and AcciMap (Wheway & Jun, 2021), or SEIPS and the London Protocol (Bourne et al., 2023).

SEIPS can be combined with other approaches to enrich the analysis of adverse events. Leon et al. (2024) analyzed a total of 60 incident reports from two English academic hospitals (between August and October 2020); the adverse events involved anticoagulants (e.g. low-molecular weight heparin) that are known high-risk medications. Interestingly, this SEIPS analysis focused on the work system factors that facilitated anticoagulant safety, such as *technology-related factors* (e.g. clinical decision alerts embedded in the EHR) and *task-related factors* (e.g. nurse being prompted to use the correct administration of medication). The researchers were also able to identify the resilient strategies of monitoring, responding, anticipating and learning (Hollnagel, 2011; Hollnagel & Woods, 2006). This study shows how various frameworks and models (e.g. resilience engineering and SEIPS) can be combined to develop a deeper understanding of adverse events in health care.

4.8 PRINCIPLES FOR HCD AND SEIPS IN ACTION

How do we design work systems and care processes for improving health care safety? In this chapter, I describe different ways that SEIPS can be integrated at multiple

TABLE 4.3
Principles for HCD and SEIPS in Action

Principles	Description
1. Person-centeredness	• The person is at the center of the work system. • The work system and care process should be designed to support the work of the person. • The person at the center of the work system can be an individual or a group of individuals.
2. Integration of HFE knowledge	• The HFE literature provides deep and broad information on what makes a 'good' work system; this knowledge needs to be integrated with the HCD process. • HFE professionals provide explicit and tacit HFE knowledge in health care safety improvement projects.
3. Systems thinking	• SEIPS is a systems approach and core systems thinking principles need to be incorporated in the HCD process. • Configuration is a unique feature of SEIPS 2.0 that supports consideration of work system interactions. • The STEW principles should be integrated in the HCD process.
4. Multiple perspectives	• SEIPS analysis should help to understand the multiple perspectives on work system barriers and facilitators in a care process. • The HCD process should incorporate multiple stakeholders, roles and perspectives through coordination, common ground and negotiation.
5. Participation	• Participation of stakeholders is critical at all stages of the HCD process. • Workload is a barrier to participation in HCD and can be partly addressed with efficient SEIPS and HFE methods.
6. Adaptation, improvement and learning	• People develop strategies in reacting to and adapting the work system and care process. These strategies should be captured and can be input to design work system interventions. • Evaluation of work system interventions can be used as input for continuous improvement, learning and continuous design.

stages of the HCD process for analysis, design and implementation of intervention, evaluation of evaluation, and analysis of adverse events. As SEIPS is integrated in HCD, it fits the characteristics of HCD, i.e. person-centeredness, participation and systems approach (Melles et al., 2021). These characteristics are the basis of the six principles for HCD and SEIPS in action; see Table 4.3.

4.8.1 Person-Centeredness – Core of HFE Discipline

SEIPS and, in particular, the work system model have been used in many different ways. For instance, the work system model was used to describe strategies used by a health care system in Hong Kong to plan for and manage the COVID pandemic (Kwan et al., 2021). Each work system element was associated with multiple strategies that were then instantiated in multiple actions. For example, the *organization element* included the strategy of 'Lead and coordinate at management level' and the following two actions: (1) forming a top management committee to formulate overall strategies and action plans and (2) forming a task force to coordinate and monitor PPE supply and consumption. In this example, SEIPS is used as a way of organizing information regarding organizational strategies and actions for managing the COVID pandemic; this allows a systematic approach that addresses all work system elements in planning for and managing a pandemic. However, there is an important feature of SEIPS that needs to be restated. SEIPS is anchored in the HFE discipline and adopts a person-centered approach; this is critical. Therefore, the person is at the center of the work system, and the work system should be designed to support the performance and well-being of the person. The person-centered characteristic of SEIPS is critical as all activities of analysis, design, implementation and improvement should be for people, to help them do their work, to support their performance and well-being, etc. The goal of SEIPS is to improve health care safety that combines patient safety and worker safety.

4.8.1.1 Person at the Center of the Work System: Individual or Group of Individuals

The person at the center of the work system can be an individual or a group of individuals. The individual could be a health care professional, a patient or a care partner. The group of individuals could be a care team composed of health care professionals, the patient and a care partner. In the SEIPS 2.0 model (Holden et al., 2013), we distinguish between three types of work system with different levels of involvement of the patient and health care professionals:

- a work system focused on the patient,
- a work system focused on the health care professional and
- a work system focused on collaborative work.

The case of collaborative work is an example where the center of the work system includes a group of individuals, i.e. health care professionals and the patient (and/or care partner).

4.8.1.2 Collaborative Work Between Patients and Health Care Workers

Collaborative work between patients and health care workers is critical for many health care safety problems, such as diagnostic safety. The 2015 IOM report on diagnostic safety dedicated an entire chapter to the concept of diagnostic team (Balogh et al., 2015). The diagnostic team includes patients, their care partners and all health care professionals involved in the patient's care. The partnership between the patient and primary care is a core component of the diagnostic team. Other health care professionals may also be part of the diagnostic team, such as radiologists, pathologists, specialists (e.g. oncologists) and others (e.g. occupational and physical therapists). The IOM conceptual model describes the diagnostic process as occurring within a work system with the diagnostic team at the center.

Some SEIPS-based methods, such as the role network analysis (Hundt et al., 2017), can help to understand and analyze the work system of teams such as diagnostic teams. The diagnostic team evolves during the diagnostic process; it is not fixed, it is fluid. Therefore, interfaces between the different work systems and changing team members need to be well-managed; this requires coordination during the diagnostic process. Any patient journey, such as the diagnostic process, can be analyzed using the SEIPS 3.0 model (Carayon, Wooldridge, et al., 2020).

4.8.2 Integrating HFE Knowledge

In sections of this chapter and Chapter 3, I discuss the use of SEIPS and HFE knowledge to design work system interventions. HFE design principles can be categorized as physical, cognitive and organizational HFE; see Section 4.5.3. The HFE literature has developed design principles for each category, such as usability heuristics for the design of health IT (category of cognitive HFE). Other disciplines have developed principles for designing 'good' jobs, such as the literature in job/organizational design and job stress. For instance, in Chapter 3, I list various work system factors that can enrich and improve work systems; see Table 3.5. This broad knowledge on what makes for a good work system needs to be integrated with SEIPS to design work system interventions. Hiring HFE professionals, consultants or experts is an important approach for integrating HFE knowledge and knowledge from other disciplines in SEIPS and HCD.

In a 2010 paper, I use the model of Greenhalgh et al. (2004) to describe HFE in health care safety as an innovation (Carayon, 2010). Many factors related to the structure and culture of health care organizations and the characteristics of their external environment can influence the adoption, implementation and sustainability of HFE (i.e. the innovation) in health care. The integration of HFE in health care organizations can take three forms (Carayon, 2010):

1. use of an HFE tool or method (e.g. conducting a SEIPS analysis of a high-risk care process),
2. training of health care staff in HFE (e.g. training staff in SEIPS and risk management in HFE) and
3. hiring of an HFE professional.

These three approaches represent a continuum where HFE knowledge is incrementally integrated in a health care organization.

Given the breadth of HFE knowledge, health care organizations should aim to use HFE tools, methods, knowledge and professionals in multiple, diverse projects, such as renovating a hospital unit, purchasing a medical device, implementing a new health IT and upgrading an adverse event reporting system. HFE professionals and experts can have different roles in health care safety improvement projects, including leaders of the projects. In Chapter 5, I describe three projects aimed at improving health care safety in various domains. HFE professionals were active participants, facilitators and leaders in the projects as they led and managed the HCD processes. Their HFE knowledge was explicitly and implicitly involved in the projects and in the design, implementation and evaluation of the work system interventions.

4.8.3 Systems Thinking

SEIPS is a systems approach and, therefore, incorporates systems thinking and principles of complex adaptive systems. The work system model is a core component of the SEIPS family; it includes multiple elements that are often represented as boxes. It is important to understand the 'boxes', i.e. all elements of the work system. The various applications of SEIPS described in this chapter show how all of the work system elements can be addressed. However, it is not just the 'boxes' (elements) of the work system that need attention; we need to pay attention to interactions between work system elements. Interconnectedness is a key feature of systems thinking that needs to be fully integrated in any application of SEIPS. Configuration is one way of ensuring consideration for interactions between work system elements.

Configuration is a unique feature of the SEIPS 2.0 model (Holden et al., 2013), which emphasizes interactions between work system elements. A configuration analysis allows the identification of a small subset of system interactions that influence outcomes. A configuration analysis of primary care nursing identified the specific work system elements involved in two processes: managing patient calls and managing walk-in patients (Tarver et al., 2024). In both processes, the following *task* factors were involved: complex patient cases needing more time and variation in day-to-day load. Unique work system factors were involved in the two processes. For instance, managing walk-in patients could be challenging depending on the location (*physical environment element of the work system*), e.g. meeting patients in the waiting area. For the management of patient calls, the sharing of workrooms could increase distractions and noise, thereby affecting the quality of phone communication. Configuration analysis has also been used to analyze interactions between work system barriers and facilitators in care transitions of older adults between the ED and home (Werner, Rutkowski, et al., 2020) and for informal care partners of people living with dementia (Weiler et al., 2022) and to compare the perspectives of hospital staff and community staff regarding the hospital discharge process (Carman et al., 2021).

Systems thinking in SEIPS can be enhanced by the integration of the STEW (Systems Thinking for Everyday Work) principles; see Table 2.2 in Chapter 2. One of the core STEW principles is to 'seek multiple perspectives'. Integrating multiple perspectives in the HCD process is critical to develop ideas for work system redesign that are relevant, feasible and accepted.

4.8.4 Multiple Perspectives in the HCD Process

Care processes typically involve numerous roles and work systems. How do people involved in a care process perceive the pluses and minuses in the care process? A few studies have compared multiple perspectives on work system barriers and facilitators. Shannon et al. (2023) conducted a study of a technologically enabled pillbox provided to patients at the time of their discharge from the hospital. Using a SEIPS-based interview guide, they interviewed six patients, two care partners and seven inpatient and outpatient clinicians about barriers and facilitators of the implementation of the pillbox. Barriers and facilitators were reported for the work system elements or a combination of work system elements. For instance, patients and care partners appreciated the *technological features* of the pillbox (e.g. useful alarms), whereas clinicians reported barriers in the *task and organization elements* (e.g. insufficient training, rushed nature of discharge). Wust et al. (2025) compared the perspectives of patients and ED nurses on work system barriers and facilitators to collaboration during ED discharge. A qualitative data analysis used interviews of 15 patients and 10 ED nurses and produced 16 dimensions of barriers and facilitators in the work system elements and combinations of elements. Many barriers and facilitators were common to both patients and ED nurses, such as the quality and content of communication between patients and ED nurses (*interaction between task and organization elements*). Patients and ED nurses also reported unique barriers and facilitators. For instance, patients reported a lack of care continuity as an *organizational barrier* when they interacted with different ED nurses at the time of discharge because of a handoff or shift change. Finally, patients and ED nurses had divergent opinions on a few work system barriers and facilitators. For instance, patients liked the flow of information on the discharge instructions document (*technology element*), whereas some nurses thought that information on how to take medications or on the diagnosis and care instructions should appear on the first page of the document.

4.8.4.1 Visualizing Multiple Perspectives on the Work System

A few studies use innovative visualizations to display multiple perspectives on work system barriers and facilitators. Goyal et al. (2020) examined patients' and nurses' perspectives regarding a bedside whiteboard for communication in hospitals. Whiteboards located in hospital patient rooms display information shared among patients, families and health care professionals. Qualitative data were collected via a patient survey (N=438) and two focus groups with 13 nurses. Barriers to whiteboard use reported by patients and nurses were displayed on the work system model and its five elements (i.e. *person, tools, tasks, environment and organization*). The work system elements included between three and five barriers and each barrier was associated with a label indicating who reported it (i.e. P for patient and N for nurse). This visual representation of barriers associated with the use of the whiteboard provided a nice overview of the perspectives of patients and nurses. Many barriers were common to both groups (e.g. abbreviations hard to understand, lack of updated clinical information from physicians) and others were unique to patients (e.g. size too small) or nurses (e.g. nurses not trained on whiteboard use).

As indicated earlier, configural diagrams are a useful tool to represent work system interactions. In a study of care transitions between the hospital and the community, Carman et al. (2021) created two configural diagrams to represent the perspectives of hospital and community health care staff on the discharge process. This visualization allows for a comparison of work system factors that are relevant from the perspectives of the hospital and community; this facilitates the comparative analysis of the complex web of work system barriers and facilitators experienced by the two groups. For instance, a barrier in one work system (e.g. missing information) led to a strategy in the other work system to facilitate work (e.g. independently gathering information). The comparison of the two perspectives also showed that a barrier in one work system could be a facilitator in the other work system. Cut-off times in a long-term care facility were seen as a barrier by the hospital staff as it may delay a patient's discharge; it was seen as a work scheduling facilitator by the community health care staff. Understanding multiple perspectives, including potential conflicting opinions, is important to improve health care safety for all involved.

4.8.4.2 Integrating Multiple Perspectives in HCD

In Chapter 1 on Health Care Safety, I describe a series of studies on HAIs conducted by our interdisciplinary team, including a study that compares work system barriers and facilitators experienced by physicians, nurses and environmental staff regarding a CDI bundle (Musuuza et al., 2019). This study provides data on common, unique and conflicting barriers reported by the three groups. The three groups did not have the same opinion on what worked and what did not work with the CDI bundle. Therefore, in subsequent steps of improving the CDI bundle, the different perspectives need to be recognized and integrated in the next iteration of the design of the bundle. Therefore, HCD projects aimed at improving care processes and health care safety need to integrate multiple perspectives.

Based on her model of collaborative design, Détienne (2006) proposed the following processes to integrate multiple perspectives in HCD projects:

- establishment of common ground,
- perspective clarification and
- negotiation with convergence and divergence.

We used Détienne's model in an analysis of the co-design process used to develop a team health IT for supporting team cognition in care transitions of pediatric trauma patients (Hose, Carayon, Hoonakker, Ross, et al., 2023). This HCD process involved multiple roles relevant to the design of the technology, including ED physician, surgeon, anesthesiologist, pediatric intensivist, nursing and pediatric trauma program management. During the co-design sessions, common ground was established through discussion of information elements that needed to be included in the technology. The discussion also allowed for perspective clarification as clinicians explained the inclusion of information elements for their own role or sometimes for the role of other clinicians. Negotiation could lead to a quick agreement after a discussion during a co-design session, but could also take time and involve a lot of back-and-forth across multiple co-design sessions. More information on this project can be found in case study #2 in Chapter 5. The mechanisms of common ground, perspective clarification and negotiation should be incorporated in HCD projects to support the integration of multiple perspectives in designing a health care safety intervention.

4.8.5 Participation in All Stages of the HCD Process

Participation is a key principle in SEIPS-based HCD processes. Stakeholders should participate in all stages and aspects of the HCD process, i.e. data collection, data analysis, design of intervention, implementation of intervention, evaluation of intervention, and continuous design of the work system change. As described in Section 4.2, participation can take multiple forms, including direct and indirect participation.

Direct participation takes place when, for instance, health care staff or patients are directly involved in designing an intervention during design sessions. Indirect participation or representative participation involves a select group of stakeholders who participate in the HCD process and represent their groups; for instance, a nurse from a hospital unit participates in workshops to design a work system redesign and represents other nurses from her unit. Her role would be to represent the opinions and views of her colleagues, to be a liaison between the unit's nurses and the design process and to share information about the design process with her colleagues. The 'selected' nurse is directly participating in HCD, whereas her colleagues are indirectly participating in HCD. Participation should also be genuine as HCD provides opportunities for stakeholders to express their opinions and have some control over the HCD process and its outcome (Østergaard et al., 2018).

4.8.5.1 Participatory HCD Process and Participatory Design Result

There is a distinction between a participatory HCD process and a participatory design result. It is important that stakeholders participate in different ways in all stages and aspects of the HCD process; this is a participatory HCD process. We also need to consider whether the end result, i.e. the work system intervention, is itself a participatory result. The end result is a participatory result when stakeholders contribute in creating choices and have decision-making power (Bratteteig & Wagner, 2016). The following types of HCD support genuine participation with stakeholder decision-making power in designing the work system intervention: collaborative design (co-design), participatory design and participatory ergonomics. The case studies described in Chapter 5 are based on co-design processes and provide examples of participatory HCD processes with work system interventions that are participatory results.

4.8.5.2 Challenges of Participation in HCD

Participation can be challenging, in particular when health care staff, patients and care partners may not be available for extended engagement in HCD. High workload experienced by health care staff is not only a major source of stress and can affect both worker safety and patient safety, but it is also a major obstacle to direct participation of health care staff in SEIPS and HCD efforts to improve health care safety. As stated by McCulloch (2024), 'The focus of our efforts needs to be directed far more towards finding out how to engage frontline staff actively in change management, since without their support, nothing works' (page 3). We need to directly address the high workload experienced by both health care workers and patients through work system redesign. We also need to create efficient SEIPS and HFE methods that engage stakeholders in various activities and support their genuine participation in HCD.

PRA or Proactive Risk Assessment is a useful approach to identify potential safety problems with an intervention, such as health IT (Carayon et al., 2011; Hoonakker, Carayon, et al., 2019); however, it has been criticized for its significant investment of resources and time necessary. In collaboration with a health care organization, we developed an efficient PRA method aimed at identifying vulnerabilities in the design and implementation of CPOE (Hundt et al., 2013). The PRA was conducted 8 months before the CPOE implementation and was conducted over three phases:

1. planning (months 1–3),
2. PRA meeting (a 5-hour meeting in month 4) and
3. short-term (month 4) and long-term (month 12) follow-up and evaluation.

The PRA focused on the high-risk transition of patients from the OR to the ICU; a new CPOE-based workflow was analyzed and evaluated by the PRA team during a 5-hour meeting. The PRA team was composed of 18 people representing nursing staff and physicians from the OR and the ICU, unit clerks and IT staff. Two HFE professionals co-facilitated the PRA team meeting, and the chief medical information officer was an ex-officio member. The PRA was deemed successful as it identified key vulnerabilities in the CPOE workflow that were addressed either before or right after CPOE implementation. The PRA was perceived as useful by the participating team members. This example shows that we can invent efficient participatory SEIPS and HFE methods that actively and directly engage busy health care staff in designing work system interventions. Virtual methods may also provide efficient ways of engaging patients, care partners and health care staff as location and schedules may be more easily accommodated (Jolliff et al., 2024).

4.8.6 Feedback Loops in SEIPS: Adaptation, Improvement and Learning

The feedback loops are a core component of SEIPS. They represent mechanisms of adaptation, improvement and learning.

4.8.6.1 Adaptation Strategies

People are at the center of the work system and interact with all other elements of the work system. Some of these interactions represent adaptive strategies where the individuals react to their experience of the work system. These strategies could be resilient strategies as discussed in Section 1.1.7 on Safety-II in Chapter 1; i.e. monitoring, responding, anticipating and learning in the context of disturbances (Hollnagel, 2011; Hollnagel & Woods, 2006). Strategies also represent behaviors of individuals who experience work system barriers. When facing work system barriers, health care workers and patients develop strategies to deal with the barriers, including working around them and delaying an action. These strategies as reactions to work system barriers have been studied, for example, among tele-ICU nurses (Hoonakker & Carayon, 2018) and care coordinators (Carayon et al., 2019).

The adaptive strategies developed and used by people can address different system levels, e.g. individual, team and organization (Li et al., 2014). Individual strategies are likely to remain at that level as the problems or barriers are probably not communicated and shared with management. This is similar to the first-order problem solving observed by Tucker et al. (2002) in a study of hospital nurses. Team-level strategies involve collaborative work and seeking help from co-workers. Organizational strategies include reporting the barriers to one's supervisor or manager. The strategies at the various levels can provide ideas for improvement and input for future work system redesigns.

4.8.6.2 Improvement and Learning in HCD

When implementing a work system intervention, we need to ask whether the change is making any difference in health care safety. This can be done by using the SEIPS model for evaluation; see Section 4.6 on SEIPS for evaluation. With the use of SEIPS and HCD, a lot is done before the intervention is implemented to optimize its fit with the work system and the care process. This anticipatory design is necessary, but not sufficient. We should expect surprises; things could happen that have not been addressed during the HCD process. Therefore, improvement should be continuous and involve activities aimed at adapting and continually improving the design of work systems and care processes with the aim of enhancing health care safety.

The third mechanism of the feedback loops in SEIPS is learning. Learning can be individual or organizational. Individual learning is important in health care as workers learn and develop skills and knowledge and patients develop strategies for dealing with their health problems. When individual learning also involves small groups of people and collaboration, then organizational learning may occur. Organizational learning is 'the process within the organization by which knowledge about action-outcome relationships and the effect of the environment on these relationships is developed' (Duncan & Weiss, 1979, p. 84). In the context of health care safety, organizational learning is important not only after investigations of adverse events but also to understand 'everyday work' and 'encourage open dialogue and collective reflection about safety' (Sujan et al., 2025, p. 143).

Organizational learning is also important in health care safety improvement projects because of the complexity of the work system and care process and the scope of the projects that typically involves multiple individuals, disciplines, perspectives, cultures and potentially organizations and settings. Lessons learned about the work system interventions and their impact need to be captured and shared throughout the organization. Lessons learned about the HCD process should also be captured and disseminated to improve future improvement projects. Organizational learning should involve applying the lessons learned to future activities and work system redesign. This is the essence of the continuous design process.

5 Case Studies and Examples of SEIPS in Action

In this chapter, I describe three selected projects to show how we have used SEIPS to improve health care safety in various domains, i.e. family-centered care and patient safety in a pediatric hospital, care transitions for pediatric trauma patients and communication and coordination in the ED discharge process for older adults. These case studies are based on action-research projects conducted in partnership with various health care organizations and multidisciplinary teams. For each case study, I define the health care safety problem addressed in the project. I describe the HCD process and the various activities performed at each stage of the process, including the intervention that was implemented and its evaluation. I explain how SEIPS was used in each project. It is important to understand that other HFE methods and models were also used in each project.

These projects use a HCD process that fits the co-design approach (Détienne, 2006). People at the sharp end of the care process (e.g. physician, nurse, patient) were directly involved in the design and implementation of interventions. A core element of the HCD process was a series of co-design sessions that were facilitated by HFE professionals. Several people representing key stakeholders and user groups of the care process applied their multiple perspectives and actively participated in the co-design sessions to develop interventions. The overall HCD process used in these projects is adapted from the ISO model (International Organization for Standardization (ISO), 2010) and includes the following four phases:

1. work system analysis,
2. specifying requirements and producing solutions,
3. co-design sessions and
4. feedback and evaluation.

I would like to thank the multiple health care organizations, collaborators and students involved in these projects. The US Agency for Healthcare Research and Quality (AHRQ) provided partial funding for these projects. The content provided in this chapter is solely my responsibility and does not necessarily represent the official view of AHRQ and other funding agencies.

 DOI: 10.1201/9781003480310-5

5.1 CASE STUDY #1 – FAMILY-CENTERED ROUNDS

In this first project, we aimed at improving the bedside rounding process for hospitalized children to make it more family-centered and further engage families. Bedside rounds allow an interdisciplinary team to interact with hospitalized patients at their bedside and review their care and treatment and other clinical decisions (Linfors & Neelon, 1980). In our project, a primary outcome of interest was patient safety, in particular medication safety: by increasing family engagement in bedside rounds, we hope to enhance opportunities for communication and coordination among the members of the care team (including parents and children), in particular, discussion about the medication regimen. In addition, we implemented an HCD process that aimed to increase the acceptance of the intervention by health care professionals with a particular focus on its efficiency; this was important given the high workload experienced by health care staff. In line with SEIPS, the bedside rounding process was embedded in a work system that we extensively analyzed before and during the design of the intervention and after its implementation.

5.1.1 Problem Addressed

Patient safety is a major issue for children (Walsh et al., 2014). A study by Kaushal et al. (2001) provides evidence of medication safety problems for hospitalized children. Using the methodology developed by Bates et al. (1995), Kaushal and her colleagues (2001) conducted a prospective study of 1,120 patients at two hospitals, and found 5.7% medication errors, 1.1% potential ADEs and 0.24% ADEs in a total of 10,778 medication orders. A recent study in two Japanese hospitals assessed adverse events among 1,126 pediatric patients (Sakuma et al., 2024). A total of 76 adverse events per 1,000 patient-days was found; 23% of them were preventable. The role of parents in ensuring the safety of the care provided to their children, in particular medication safety, has long been recognized (Mazor et al., 2010). In this project, our aim was to increase engagement of families in the care of their hospitalized children, in particular during the rounding process. Well-designed bedside rounds can actively engage children and their families. This provides a unique opportunity for improving communication and coordination of critical information for the safe care of children; therefore, increasing patient safety and, in particular, medication safety (Muething et al., 2007).

5.1.2 Human-Centered Design Process

The HCD process began with a deep analysis of the bedside rounding process with multiple data collection methods, i.e. observation and videorecording of rounds, and interviews of health care professionals and parents using the stimulated recall method (Carayon Li, et al., 2014). All of these data were used to identify ideas and strategies for improving family engagement in rounds (Kelly et al., 2013). We also managed a series of five co-design sessions with a 15-member multidisciplinary team (Xie, Carayon, Cartmill, et al., 2015). The multidisciplinary team consisted of one parent from the hospital's patient and family advisory council, three senior physicians, two resident physicians, a medical administrator of the hospital, two nurse managers, two nurses and four HFE professionals. During the HCD process, additional input from health care professionals and parents was gathered via multiple surveys, such as a survey on the potential safety impact of identified strategies (Xie et al., 2012). An intervention was designed, pilot tested, refined and finally implemented and evaluated in a pediatric hospital (Cox et al., 2017). The intervention consisted of a checklist of eight practices for engaging families in rounds, e.g. 'discuss assessment & plan for day with family'. The checklist also included tips for each practice; e.g. 'use lay terms & trends; make eye contact with family & answer questions'. Its implementation was facilitated by multiple training sessions with role-playing and informational meetings. The following HFE principles were integrated in the design of the intervention: shared mental model, usability, workload consideration and systems approach (Xie, Carayon, Cox et al., 2015). In particular, the design of the checklist benefited from an extensive review of HFE research on checklists; see the Appendix in Xie, Carayon, et al. (2015). For more information on the HCD process used in this project, see Xie, Carayon, Cartmill, et al. (2015) and Xie, Carayon, Cox, et al. (2015).

In a cluster randomized trial with 298 families (Cox et al., 2017), we found that increased use of the checklist was associated with higher family engagement (as observed in videorecording of rounds) and more positive parent perceptions of safety climate (as measured by a survey). Some of the checklist items were particularly important: the health care team reading back orders and the health care team asking the family for questions. Resources developed and used in this project are described in a toolkit available at: https://hipxchange.org/toolkit/familyrounds/.

5.1.3 How was SEIPS Used?

SEIPS was used in several stages of the HCD process:

- description of the bedside rounding process (Xie, Carayon, Cox, et al., 2015)
- analysis of work system barriers and facilitators in the rounding process using a SEIPS-based interview guide and the stimulated recall method (Carayon, Li, Kelly, et al., 2014)
- identification of potential solutions mapped onto the work system (Kelly et al., 2013)
- pilot study of the intervention: identifying work system pluses and minuses of the intervention and using this information to improve the design of the intervention before its full implementation (Xie, Carayon, Cox, et al., 2015)
- evaluation of the implementation of the intervention: using SEIPS-based interview guide to identify work system barriers and facilitators associated with the intervention (Kelly et al., 2019).

In the initial phase of HCD, we conducted more than 60 hours of observation of the bedside rounding process, as well as preparation to and follow-up of rounding, in various units of the participating pediatric hospital. This allowed us to become familiar with the rounding process and to create a map of the process (Xie, Carayon, Cox, et al., 2015).

The process map shows a series of steps, including before round, during round and after round, with different *individuals* performing *tasks* in a *physical environment*. For instance, the round itself is performed either in the patient room or in the hallway, depending on the particular situation of the child (*work system element of physical environment*). In line with SEIPS 2.0 (Holden et al., 2013), this SEIPS-based process map shows the collaborative work done by hospital staff, parents and children.

To go beyond the description of the rounding process and to understand work system barriers and facilitators to family engagement in the rounding process, we conducted stimulated recall interviews (Carayon, Li, Kelly, et al., 2014). The initial stimulated recall interviews were conducted with five health care team members (two senior physicians, two junior physicians and one nurse) and five parents. The participants were first asked to review the video recording of their own rounding session. We then discussed work system barriers and facilitators of family engagement evident in the video. Interview probes covered all of the work system elements. The deductive analysis produced a range of barriers and facilitators in each of the work system elements: e.g., the team listening to the family was an *organization* facilitator and a lack of clarity in assessment and plan was a *task* barrier; see Figure 5.1. Some of these barriers and facilitators were specifically addressed by the intervention. For instance, the checklist included the items of 'discuss assessment and plan for day with family' and 'ask family for questions'.

Additional stimulated recall interviews were conducted to identify potential solutions or strategies for increasing family engagement during rounds (Kelly et al., 2013). A total of 37 interviews with 11 parents, four children (aged 8–17 years) and 22 health care team members (eight senior physicians, six junior physicians, five medical students and three nurses) produced a range of solutions for the phases of before, during and after the round. For each work system element, we identified strategies, such as clearly defining the roles of health care team members during the round (*interaction between task and organization elements*). The multidisciplinary team discussed these solutions and strategies during the co-design sessions and decided to focus on what happens during the round itself. A checklist of good practices was then developed.

We developed different prototypes of the checklist with input and feedback from the multidisciplinary team during the co-design sessions. We conducted a pilot study of a full prototype of the intervention in two hospital units and observed 47 rounds (Xie, Carayon, Cox et al., 2015). We found barriers and facilitators in the use of the checklist for all work system elements. For instance, in one of the units, the health care team would bring a workstation-on-wheels to the round; so, we adapted the intervention to fit the *technology* (i.e. workstation-on-wheels). In the unit where the team used the workstation-on-wheels, we attached a plasticized version of the checklist to the workstation. The pilot study produced important information used to adapt the intervention before its full implementation and evaluation.

Once the intervention was implemented, we went back and observed the rounding process again. We observed a total of 88 rounding sessions and interviewed seven participating physicians (Kelly et al., 2019). We found many facilitators to the use of the checklist; e.g. familiarity of the health care team with the checklist (*person*) and

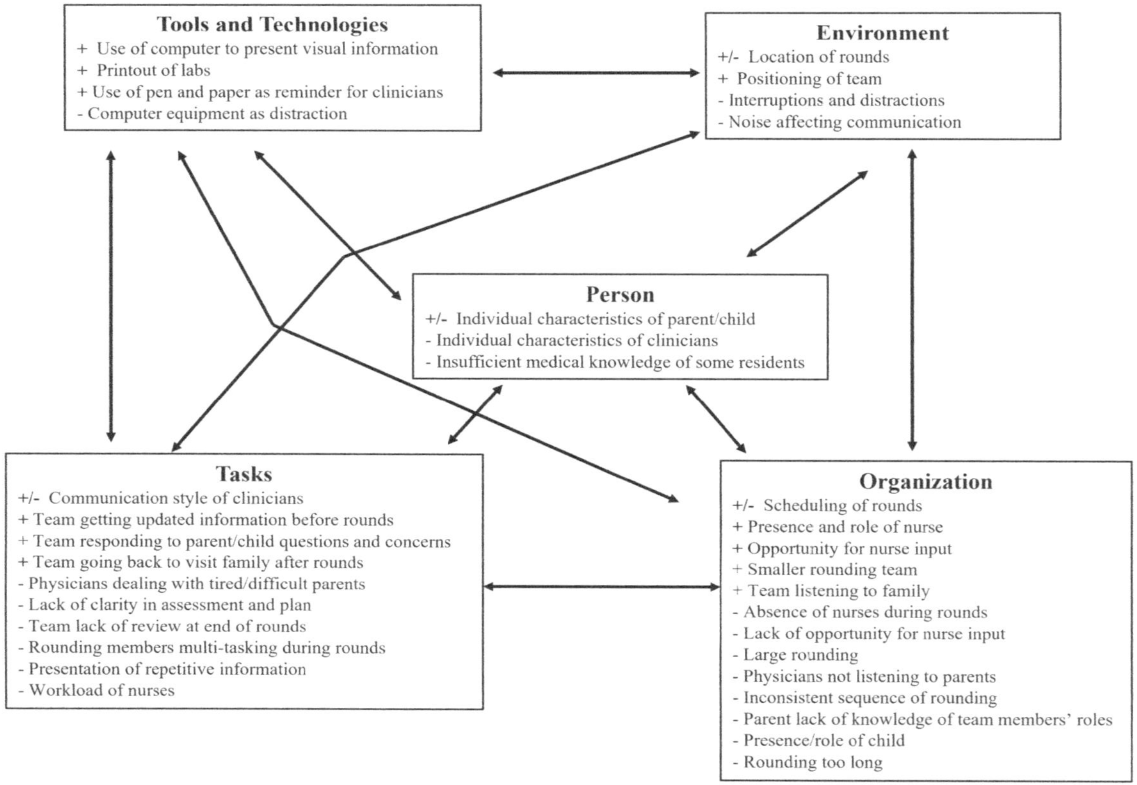

FIGURE 5.1 Work System Barriers and Facilitators to Family Engagement in Rounds (Carayon, Li et al., 2014) [Note: The '+' symbol indicates a facilitator and the '–' symbol indicates a barrier. Note that barriers and facilitators are assigned to one work system element; however, many system elements interact with each other.]

visibility of the checklist (*technology*). But we also found barriers, such as the high workload of the health care team (*task*) and a lack of presence of parents (*organization*). Some of these barriers were not addressed by the intervention, but need to be considered in subsequent design steps.

In this project, the HCD process included multiple methods, steps and iterations with input from many users and stakeholders. Overall, the intervention had a positive impact on family engagement. Postimplementation, we also identified both barriers and facilitators to the intervention. This clearly shows challenges in designing interventions; therefore, calling for a continuous design and improvement process.

5.2 CASE STUDY #2 – TEAM HEALTH IT FOR CARE TRANSITIONS OF PEDIATRIC TRAUMA PATIENTS

In the second project, we aim at improving communication and coordination in care transitions of pediatric trauma patients between the ED, OR and the PICU. A primary outcome of interest was patient safety, in particular avoiding safety problems associated with missing or inadequate patient-related information and ensuring that all critical injuries are identified and treated (Hoonakker, Wooldridge, et al., 2019). In addition, given the high workload in the ED, OR and PICU, the intervention was designed to be efficient so that it would be accepted by all health care staff involved in the care transitions. In line with the SEIPS 3.0 model (Carayon, Wooldridge, et al., 2020), the care transitions were embedded in multiple work systems that we extensively analyzed during the HCD process.

5.2.1 Problem Addressed

Care transitions are often described as contributing to safety problems. When a patient transitions between care settings, such as between hospital units or between a hospital and a long-term care facility, critical information may not flow effectively and efficiently, and miscommunications may occur; this then leads to adverse events. However, it is important to recognize that care transitions also represent opportunities for improving patient safety. For instance, a second pair of eyes (e.g. a physician in the receiving facility or unit) reviews what is happening with the patient and may identify issues with the care pathway (Wears et al., 2012). Therefore, care transitions are critical moments for safety in the patient journey.

In this project, we examine care transitions experienced by pediatric trauma patients. When a child is injured in an accident and goes to the ED because of a trauma, a large number of health care professionals may be involved, especially as the child needs surgery and is transferred to the OR and/or the child is critically injured and is transferred to the PICU. In an analysis of a pediatric trauma registry, we identified 53 roles involved in the pediatric trauma care process, e.g. physicians in different specialties (e.g. emergency medicine, surgery), nurses and pharmacists (Wooldridge et al., 2019). Ensuring effective and timely communication and coordination among members of such a large, distributed health care team is challenging and can have a major impact on the quality and safety of care for pediatric trauma patients (Hoonakker, Wooldridge, et al., 2019). In the project, we aimed to design a health IT to provide cognitive support to this large, distributed health care team involved in the care transitions between the ED, OR and PICU.

5.2.2 Human-Centered Design Process

We began the HCD process with interviews of clinicians from the ED, OR and PICU to understand care transitions between the ED, the OR and the PICU, and barriers and facilitators to communication and coordination during the transitions (Wooldridge et al., 2020). These data were also used to generate potential solutions, which were discussed by an 11-member multidisciplinary team. The multidisciplinary team consisted of one ED physician, one pediatric surgeon, one pediatric anesthesiologist, one pediatric intensivist, one pediatric hospitalist, the chief medical information officer, the pediatric trauma nurse program manager and four HFE professionals. The team met in a total of four meetings or co-design sessions over 10 months (Carayon et al., 2022). The team decided to focus on developing a team health IT, which we called T^3 or Teamwork Transition Technology; the aim of T^3 was to provide cognitive support to the large, distributed team. The design of T^3 integrated the following HFE design principles: situation awareness, physical and visual access, visual support, information integration and interpretation, consistency, minimization of workload and safety (Carayon et al., 2022).

Because a team health IT such as T^3 entails meeting the information needs of various people and disciplines (e.g. emergency medicine, trauma surgery, nursing), our work during and in between co-design sessions involved the identification of relevant information elements and the selection of key elements to display on T^3. A major challenge of this co-design process was the management of multiple perspectives (Hose, Carayon, Hoonakker, Ross, et al., 2023). Multiple versions of T^3 were created and iteratively improved with input from the multidisciplinary team and other stakeholders; see Hose, Carayon, Hoonakker, Ross, et al. (2023) for the different prototypes and the final mock-up of T^3.

We conducted a scenario-based evaluation of T^3 with 36 health care staff in various units and services (Hoonakker et al., 2022). This evaluation showed that T^3 provided cognitive support to health care staff involved in pediatric trauma care transitions (Hoonakker et al., 2022) and was highly usable (Carayon et al., 2022). For more information on the HCD process used in this project, see Carayon et al. (2022).

5.2.3 How was SEIPS Used?

SEIPS was used in several stages of the HCD process:

- analysis of work system barriers and facilitators in care transitions between the ED, OR and PICU using a SEIPS-based interview guide (Wooldridge et al., 2020)
- description of ED-OR-PICU care transitions using the SEIPS-based process modeling method (Wooldridge et al., 2022)
- proactive risk assessment (PRA) using the work system model to identify what can go wrong with T^3, safety consequences and what can be done to address the vulnerabilities (Hoonakker, Carayon, et al., 2019)

- scenario-based evaluation of T^3 mock-up (Hoonakker et al., 2022): using SEIPS-based interview guide and identifying work system barriers and facilitators of T^3 (Hose, Carayon, Hoonakker, Brazelton, et al., 2023).

The initial SEIPS analysis consisted of interviewing 18 senior and junior physicians and nurses involved in care transitions between the ED, OR and PICU (Wooldridge et al., 2020). We asked each interviewee to describe in detail the care transitions in which they participate as well as 'bad' and 'good' care transitions, i.e. transitions where care for the pediatric trauma patient was or was not compromised. We asked follow-up and probing questions based on the work system elements. The qualitative data analysis of the interviews produced nine themes or dimensions of work system barriers and facilitators: anticipation, ED decision-making, interacting with family, physical environment, role ambiguity, staffing/resources, team cognition, technology and characteristics of trauma care. We mapped the dimensions to the work system elements and their interactions. For instance, anticipation, or how health care team members foresee and address in advance the needs of children, is related to the *task and organization elements*. Based on the analysis of work system barriers and facilitators, the multidisciplinary team decided to focus on the dimension of team cognition. We developed a technology to provide cognitive support to health care staff involved in pediatric trauma care transitions. The technology was called T^3 or Teamwork Transition Technology.

The interview data were also used to create detailed SEIPS-based process maps, in particular for the OR to PICU care transition (Wooldridge et al., 2022); the SEIPS-based process modeling method is explained in Chapter 3. The maps of the care transitions provided a visual representation of the complex work in care transitions. In particular, the sometimes 'invisible' articulation work was clearly identified and described, such as preparation for a care transition (e.g. organizing transport).

Any change in a work system impacts the entire work system; it is, therefore, important to analyze the change before it is implemented in order to anticipate potential negative consequences, in particular on patient safety. PRA is a method aimed at identifying potential risks of an intervention, such as a technological change (Carayon et al., 2011). In a manner similar to the 'efficient proactive risk assessment' method that we developed and used to evaluate the impact of CPOE on the care transition of patients between the OR and the ICU (Hundt et al., 2013), we organized a two-hour PRA meeting with a pediatric ED physician, a pediatric ED nurse, a pediatric surgeon, a pediatric anesthesiologist, an OR nurse, two PICU nurses and the pediatric trauma coordinator (Hoonakker, Carayon, et al., 2019). The PRA of the T^3 mock-up addressed the following questions:

- What can go wrong with T^3? [vulnerabilities]
- What are possible safety consequences?
- Why would something go wrong with T^3?
- What can be done to address those issues?

We displayed the SEIPS model in the meeting room in order to draw participants' attention to the systems approach used in the PRA. When discussing vulnerabilities and potential solutions to address the vulnerabilities and their safety consequences, we referred to the work system model; this helped participants think about solutions related to the technology itself as well as the other work system elements. For instance, a potential risk was that patient-related information displayed on T^3 could be wrong; e.g. because of a data entry problem or a delay in displaying critical information. Solutions discussed by the multidisciplinary team addressed the *technology* itself (e.g. clear display of when the information was last updated) as well as the role of ED nurses in entering data into the EHR and T^3 (*interaction between the technology, person-nurse and task elements*).

For the evaluation of the T^3 mock-up, we worked with our clinical collaborators to develop a realistic scenario of a pediatric trauma patient and their transition from the ED to the OR and the PICU (Hoonakker et al., 2022). Using the System Usability Score (SUS), we showed that T^3 had a high usability score (Carayon et al., 2022). However, as I discussed multiple times in this book, a change in the work system impacts all parts of the system, sometimes in surprising ways and often in both positive and negative manners, even though a HCD process with integrated HFE knowledge is used. This is also the case for T^3. In the scenario-based evaluation of T^3, we asked the 36 participants to tell us what they like and what they dislike about T^3, and to explain how T^3 supports or hinders the work and workflow of caring for pediatric trauma patients, in particular during transitions between ED, OR and PICU (Hose, Carayon, Hoonakker, Brazelton, et al., 2023). We created probes for the interview questions that were based on the work system elements. The qualitative data analysis produced a total of eight dimensions organized as follows:

- *Person*: dimension of experience,
- *Task*: dimensions of task performance and workload/efficiency,
- *Technology*: dimensions of usability and specific features of T^3,
- *Environment*: dimensions of space for T^3 and location of T^3 and
- *Organization*: dimension of communication/coordination.

For each dimension, we counted the number of interviewees who mentioned a barrier or a facilitator (see Figure 5.2). In general, results about T^3 were positive. As shown in Figure 5.2, interviewees reported facilitators more often than barriers in the use of T^3. In particular, a large majority talked about T^3 as helping perform various *tasks* (e.g. understanding the child's condition, identifying what needs to be done and preparing the child for transition).

A major challenge in this project was addressing the various needs of multiple health care staff involved in care transitions of pediatric trauma patients. This management of multiple perspectives received particular attention during the co-design sessions when, for instance, different opinions were expressed that did not converge (Hose, Carayon, Hoonakker, Ross, et al., 2023). See more discussion on collaboration and multiple perspectives in HCD in Section 4.8.4 of Chapter 4.

Work system element	Dimension	Barrier definition	Facilitator definition	B (# interviewees)	F (# interviewees)
Person	Experience	*Less* clinically experienced care team members are *less* likely to understand T^3	*More* clinically experienced care team members are *more* likely to understand T^3	1	1
Task	Task performance	T^3 *hinders* team member's ability to achieve a task	T^3 *supports* team member's ability to achieve a task	4	34
Task	Workload & efficiency	T^3 *creates additional work* for team member	T^3 *reduces time or effort* for team member	5	15
Technology	Usability	The design of T^3 *affects* effectiveness, efficiency and satisfaction of users*	The design of T^3 *supports* effectiveness, efficiency and satisfaction of users*	11	22
Technology	Specific features of T^3	The design of any of the 11 specific features of T^3 is a *negative* aspect	The design of any of the 11 specific features of T^3 is a *positive* aspect	25	30
Environment	Space for T^3	There is *insufficient or inadequate room* in the physical environment to place T^3	There is *sufficient or adequate room* in the physical environment to place T^3	5	1
Environment	Location for T^3	Specific place or position in the physical environment of T^3 *hinders* its use	Specific place or position in the physical environment of T^3 *facilitates* its use	4	32
Organization	Communication & coordination	T^3 *hinders* exchange of information and coordination between care team members~	T^3 *supports* exchange of information and coordination between care team members~	1	32

Note: total number of interviewees = 36

FIGURE 5.2 Work System Barriers and Facilitators Associated with T^3 [Teamwork Transition Technology] (Hose, Carayon, Hoonakker, Brazelton, et al., 2023).

5.3 CASE STUDY #3 – IMPROVING ED DISCHARGE PROCESS FOR OLDER ADULTS

In the third project, we aimed at improving the ED discharge process for older adults. The ED discharge process begins in the ED itself and is extended to follow-up by ED nurses once the patient is at home. In a manner similar to the other two projects, we used an HCD process that addressed the needs of both patients and health care staff. We aimed to enhance patient engagement in their care and improve patient safety, such as ensuring timely and adequate follow-up care (e.g. visit with the primary care physician or a specialist) and treatment (e.g. new medications prescribed in the ED). Given the high workload in the ED, we paid attention to the efficiency of the intervention with an effort to not add burden to the busy ED health care staff. In line with the SEIPS 3.0 model (Carayon, Wooldridge, et al., 2020), we extensively analyzed the multiple work systems in the ED (e.g. ED physicians and nurses) and the work system of older adults; these work systems are interconnected as they are involved in the extended ED discharge process.

5.3.1 Problem Addressed

Older adults (aged over 65) are at high risk of adverse events, such as falls, functional decline and HAIs (Long & Vincent, 2012). Care transitions are particularly vulnerable moments in the care journey of older adults (Arbaje et al., 2014), especially when they transition through the ED (Shah et al., 2018). After an ED visit, older adults are at high risk of adverse events, such as readmission to the ED or unplanned hospitalization (Schnitker et al., 2011). Older adults represent a disproportionately high number of ED visits, i.e. 19% of ED visits in the United States, for a total of 27 million visits (Cairns & Kang, 2021). As explained in Case Study #2, care transitions can be sources of safety problems but can also offer opportunities for enhancing patient safety. Care transitions from the ED to home are particularly complex (Werner, Rutkowski, et al., 2020) and are in need of improvement to ensure high quality and safety of older adults' care. Interventions to improve the ED discharge process for older adults have been narrow and focused. In this project, we designed, implemented and evaluated a patient-centered intervention for improving communication and coordination during the extended ED discharge process for older adults (Carayon, Wust, et al., 2025).

5.3.2 Human-Centered Design Process

We began the project with extensive data collection to understand older adults' experience in the ED as well as the viewpoint of health care staff in the ED, hospital, primary care and long-term care regarding care transitions for older adults (Werner et al., 2025). All data were analyzed and then reviewed and discussed by our large project research team. The decision was made to focus the intervention on the ED discharge process, i.e. when a patient leaves the ED and returns home. Similar to the two projects described earlier, we created a multidisciplinary design team with the objective of designing and implementing the intervention. The nine-member multidisciplinary design team consisted of one care partner, one ED physician, one ED nurse, one nursing professor with expertise in geriatrics and transitional care, two IT professionals, and three HFE professionals. Because this part of the project occurred during the COVID pandemic, the team met remotely for five 1-hour co-design sessions over five months. The intervention consisted of four components distributed in the three subprocesses of the ED discharge process (see Table 5.1):

1. We designed a new template in the EHR to help physicians write patient-centered instructions.
2. After multiple iterations and feedback from several groups of older adults and community members, and with support from the IT group, we redesigned the after-visit-summary (AVS) document that provides discharge instructions to ED patients.
3. We worked with ED nursing leadership to enhance ED nurses' training for teach-back.
4. We worked with ED nurses and the IT group to embed tips into the EHR that provided cognitive support to nurses when they called the patient at home after their ED visit.

TABLE 5.1
SEIPS Description of the Patient-Centered ED Discharge Process Redesign (Carayon, Wust, et al., 2025)

	Subprocesses		
Work System Elements	**1. Writing discharge instructions**	**2. Communicating instructions at discharge**	**3. Communicating at follow-up call**
Person	Physician	Nurse and patient	Nurse and patient
Task	Enter discharge instructions	Communicate with each other about discharge	Communicate about follow-up care
Technology	Into EHR [NEW-1.template for discharge instructions]	Using the AVS [NEW-2.redesigned AVS]	Nurse using patient-related information available in the EHR [NEW-4.tips embedded into the EHR]
Organization	Once the physician is done with writing discharge instructions, a trigger is sent to the nurse who can print the AVS.	The nurse may use teach-back to ensure patient comprehension of discharge instructions. [NEW-3.re-emphasis/ reminder about teach-back]	Nurse who calls the patient is unlikely to be the nurse who cared for the patient in the ED.
Physical environment	At a computer workstation in the ED	In the ED patient room	Nurse in the ED; patient at home

Note: AVS = after-visit-summary.

ED clinicians, patients, care partners and experts in IT, HFE and geriatrics participated in the HCD process via multiple mechanisms, including the co-design sessions, separate meetings with members of the multidisciplinary team and experts and usability evaluation of initial redesigns of the AVS (Carayon, Wust, et al., 2025; Wust et al., 2023). The intervention was implemented in the ED of a large academic hospital in the United States; initially designed for older adults, the intervention was implemented for all ED adult patients. A mixed-methods evaluation of the intervention showed that it was perceived positively by ED physicians and advanced practice providers (APPs), ED nurses and older adult patients (Carayon, Wust, et al., 2025). For more information on the HCD process, see Carayon, Wust, et al. (2025).

5.3.3 How was SEIPS Used?

SEIPS was used in several stages of the HCD process:

- analysis of work system barriers and facilitators in care transition from the ED to home for older adults, using a SEIPS-based interview guide with health care staff and contextual inquiry observations of older adults in the ED (Werner et al., 2025)
- SEIPS-based journey map of the older adult's visit in the ED (Wust et al., 2022)
- description of the intervention using the work system model; see Table 5.1 (Carayon, Wust, et al., 2025)
- evaluation of the intervention with identification of work system barriers and facilitators related to the intervention (Carayon, Wust, et al., 2025).

In a manner similar to the other two case studies, a SEIPS analysis was initially conducted to understand the care transitions of older adults from the ED to home, a hospital unit or a long-term care facility (Werner et al., 2025). Using the contextual inquiry method (Beyer & Holtzblatt, 1999), we shadowed 20 older adults after their arrival to the ED until their departure and transition to home, the hospital or long-term care (total of 103 hours of observation). We also interviewed 31 health care professionals in the ED, hospital, primary care and long-term care about their perceptions of what works and what does not work in ED care transitions for older adults (average duration of interview: 51 minutes). The data collection benefited from probes related to the work system elements. The thematic qualitative data analysis provided rich information on the ED journey for older adults. The journey consists of several subprocesses: intake and assessment, diagnostic process, disposition decision-making, discharge, coordination between ED and hospital, coordination between ED and outpatient setting, and ED follow-up with the patient at home. For each subprocess, we identified specific barriers and facilitators, such as barriers and facilitators in writing discharge instructions in the EHR. We also identified barriers and facilitators that influence all subprocesses and the entire ED journey, e.g. teamwork in the ED (barrier or facilitator) and time constraints in the ED (barrier). Following this in-depth SEIPS analysis, the project research team decided to focus on redesigning the ED discharge process.

In line with the SEIPS 3.0 model (Carayon, Wooldridge, et al., 2020), we developed the SEIPS-based patient journey mapping method and applied it to the ED patient visit (Wust et al., 2022). The method produces a visual representation of the patient's journey and all work system elements. The ED patient journey map consisted of three segments organized sequentially: (1) arrival to the ED and intake; (2) assessment, diagnosis and decision on next steps; and (3) admission to the hospital or discharge. For each segment, the map includes information on *people* doing *tasks* using various *tools and technologies* and collaboration between different people (*organization element*); the location of the tasks (*physical environment*) is also included. The patient journey map can also be used as a visual tool for explaining the ED care process to patients and care partners and helping them understand what to expect during their ED visit (Reay et al., 2017).

An intervention can have multiple components and is, of course, embedded in the work system. This is what we show in Table 5.1 (Carayon, Wust, et al., 2025). A SEIPS-based description of an intervention shows how the intervention fits in the entire work system. This can help the designers and implementers of the intervention think about the potential impact of the intervention on the work system, including its anticipated barriers and facilitators.

The evaluation of the intervention consisted of mixed methods (EHR data, survey and interview) and addressed multiple perspectives, i.e. ED physicians, APPs and nurses and older adult patients (Carayon, Wust, et al., 2025). We conducted a survey of 42 ED physicians and APPs. We also interviewed 10 ED nurses and 15 patients at home. The survey and interview qualitative data were analyzed in light of the work system model. For each of the three subprocesses targeted by the intervention (see Table 5.1; subprocesses of writing discharge instructions, communicating at discharge and communicating at follow-up), we identified barriers and facilitators for all work system elements. For instance, physicians and APPs described the EHR template for writing discharge instructions as well-designed (facilitator in the *technology element*). They also indicated that more prepopulated information could be integrated into the template in order to speed up the subprocess of writing discharge instructions (barrier in the *task element* related to the *technology element*). In general, more facilitators were mentioned than barriers, indicating an overall positive evaluation of the intervention. Both nurses and patients mentioned several positive aspects of the redesigned AVS (after-visit-summary), including its usability, layout, font size and content of information (facilitator in the *technology element*). However, the barriers identified by ED health care staff and older adults should be addressed in subsequent efforts. Information on the barriers represent opportunities for improvement and learning, i.e. the feedback loops in SEIPS.

5.4 LESSONS LEARNED AND CHALLENGES

The three projects described in this chapter provide information on different ways that SEIPS can be used in practice. These are just examples; many examples are available in the scientific and gray literatures. Many other applications remain to be invented and created.

In the three projects, SEIPS was embedded in multiple stages of the HCD process for the following activities:

- SEIPS-based analysis to describe the care process and identify work system barriers and facilitators with data collected via interview, observation, video recording and contextual inquiry method.
- SEIPS for designing the intervention; e.g. identifying potential solutions, testing a prototype, identifying safety risks of the intervention, describing the intervention.
- SEIPS for evaluating the intervention; e.g. identifying work system barriers and facilitators related to the intervention.

The three projects share common approaches in the four stages of the HCD process (Carayon, Hoonakker, and Hose, 2025):

1. Stage of work system analysis: use of multiple SEIPS-based methods to understand the care process and identify work system barriers and facilitators.
2. Stage of requirement specifications and identification of potential solutions: use of initial work system analysis to identify potential solutions, integration of multiple HFE principles in the design of the interventions and creation of prototypes with feedback from key stakeholders.
3. Stage of co-design: multidisciplinary team involved in multiple co-design sessions over a period of 5–10 months, significant preparatory work and follow-up between meetings.
4. Stage of feedback and evaluation: evaluation of intervention with multiple methods; e.g. interview, PRA and scenario-based evaluation.

The HCD process used in all three projects was based on collaborative design (Détienne, 2006) and included a multidisciplinary team who met in several design sessions over an extended period of time. The multidisciplinary teams met between four and five times over a period of 5 to 10 months. They included between nine and 15 members, including three to four HFE professionals. Different disciplines and roles were represented on the teams depending on the project's objectives. For instance, in the two projects involving the design of a health IT, people with IT expertise were included in the multidisciplinary teams. In Case Study #2, the chief medical information officer was a member of the team, and in Case Study #3, two IT professionals participated in the design team. In two projects, the team included a patient's representative. In Case Study #1, a parent whose child had been recently hospitalized was a member of the design team; she was also a member of the hospital's patient and family advisory council. In Case Study #3, the care partner of an older adult who had visited the ED multiple times within the past year participated in the design team.

Participation of key stakeholders was a key feature of the collaborative design process implemented in all three projects. Participation took multiple forms of direct and indirect participation. Direct participation occurred via membership in the multidisciplinary design teams. Indirect participation took different forms, including surveys of parents and clinicians in Case Study #1. In addition, the people who participated in the multidisciplinary teams were charged with being liaisons between the teams and their colleagues. The collaborative design process included direct participation of people at the 'sharp' end (e.g. physician, nurse, patient) as well as people at the 'blunt' end, who represented organizational leadership (e.g. nurse manager, hospital administrator) and operations (e.g. IT professionals).

The projects were managed and facilitated by HFE professionals in close collaboration with clinicians, patients and their representatives and leaders from the participating health care organizations. The projects used SEIPS-based methods in different ways and at different stages of the HCD process. HFE knowledge was formally integrated through the use of HFE methods (including SEIPS-based methods) and the application of HFE design principles, such as shared awareness, usability and minimization of workload. Tacit HFE knowledge was embedded in various discussions and activities of the HCD process through the active participation of HFE professionals. This poses the question of how to embed and integrate HFE knowledge in health care.

The three projects addressed specific health care safety problems over a limited period of time. We need to determine how to embed HFE knowledge within health care organizations to sustain HFE-based improvement over time. Podtschaske et al. (2018) described a program at Stanford Health Care that integrated clinicians with experts in safety sciences, systems engineering and HFE. SEIPS has been used in collaborative projects to address multiple health care safety problems, e.g. the safe administration of blood and blood products. Sutcliffe and Lomax (2025) describe the implementation of an HFE program within a United Kingdom hospital that includes HFE training and small- and large-scale HFE projects. SEIPS is used in HFE projects to analyze work systems and care processes, such as the escalation of deteriorating patients in the maternity unit. The US Agency for Healthcare Research and Quality (AHRQ) created the Patient Safety Learning Laboratory (PSLL) program with the goal of supporting the emergence of interdisciplinary research teams that take a systems engineering approach (e.g. SEIPS) to address health care safety problems. For more information on the AHRQ PSLL program, see: https://ahrq-stage.ahrq.gov/patient-safety/resources/learning-lab/index.html. This has facilitated the integration of HFE in health care safety research projects, such as usability of patient portal and patient safety dashboard (Businger et al., 2019). The creation of such 'interdisciplinary learning ecosystems' for health care safety requires adequate organizational design and resources, in particular various interpersonal/interprofessional, informational, structural, learning and supporting activities (Atkinson et al., 2022). Changes at multiple levels (e.g. health care organizations, policy, technology designers and vendors) are necessary to support the long-term effective and efficient integration of HFE in health care safety improvement (Carayon et al., 2018).

6 Reflections on SEIPS in Action and Looking Forward

Redesigning work systems and care processes for health care safety is difficult and requires the genuine integration of multiple perspectives and approaches, such as SEIPS. In this chapter, I explore challenges in SEIPS and HCD, including moving beyond analysis and designing work system interventions that improve health care safety. Critical questions include: how do we engage in continuous design and learning? Whose work are we redesigning? As elsewhere in this book, I navigate between conceptual, methodological and practical issues and hope to provide ideas and suggestions for future development, research and application of SEIPS.

6.1 DESIGN IS CHALLENGING; LEARNING IS ESSENTIAL

> We must accept finite disappointment, but never lose infinite hope.
>
> (Martin Luther King Jr.)

Human-centered design for health care safety is challenging. It needs to achieve multiple objectives, in particular, improvement in both patient safety and worker safety. It needs to integrate multiple perspectives and potentially deal with disagreement and conflict to achieve consensus on a work system redesign. It needs to integrate systems thinking and its various principles (see Table 2.2 on the STEW principles) and to incorporate HFE knowledge in creating the work system redesign. It needs to produce an intervention that fits the entire work system and the care process and achieve safety improvement for both patients and workers. In this section, I focus on the challenge of moving from the analysis stage to the design stage of HCD and how SEIPS and HFE are integrated in HCD. I conclude with a discussion about who the designers are.

6.1.1 From Analysis of Problem to Design of Solution(s) – A Learning Cycle

Moving from analysis to solution is not an easy process. In Chapter 4, I describe different ways that we can translate a SEIPS analysis into work system redesign. A SEIPS analysis often produces information on work system barriers and facilitators; this input can help decide what needs to be changed in the work system to eliminate or mitigate work system barriers and to keep or enhance work system

DOI: 10.1201/9781003480310-6

facilitators. A SEIPS analysis can also be complemented by a thematic analysis of work system interactions and produce design themes or requirements; see, for example, Sujan, Crumpton, et al. (2024). During a SEIPS-based analysis of a specific health care safety problem, information may also be collected about ideas for work system redesign. These ideas are then discussed, for instance, during design sessions or workshops. We also need to assess whether the potential solutions fit with the work system(s) under study; this requires a SEIPS analysis of the solutions.

Moving along the HCD process from analysis to identify potential solutions, and to decide on a specific intervention, requires imagination, creativity and openness, especially when applying a systems approach such as SEIPS; we need to consider the multiple complex work system interactions and their impact on people. What innovative mechanisms and methods can we invent to build bridges between the phases of analysis and solution design in HCD? As indicated in Chapters 4 and 5, visual methods are particularly helpful to understand system interactions and could be further expanded to support the work of going from analysis to design of a solution.

A study by Yardley et al. (2022) shows the challenge of coming up with interventions when dealing with a complex system and complex mechanisms, such as those involved in providing out-of-hours palliative care, and addressing health care safety issues (e.g. unnecessary patient and carer distress). The study uses (1) SEIPS to describe the work system and wanted/unwanted outcomes and (2) realistic approaches to identify context-mechanism-outcome (CMO) configurations. The researchers identified six CMO configurations that could be addressed with various interventions. A CMO example was the involvement of multiple care professionals (context) who used different IT (mechanism); there was also uncertainty about who to contact for what (mechanism). These challenges were related to interactions among multiple work system elements: that is, *technology and tools* (IT), *organization* (multiple care professionals, responsibilities), and *external environment* (national policies). The challenges resulted in the following outcomes: untimely access to patient records and potential suboptimal care because of incomplete information. A technological solution to support care coordination among non-co-located individuals working in different organizations (e.g. technological interfaces to allow access to patient-related information in a timely manner) would help to support information flow. However, the technological solution would not be sufficient; attention needs to be given to the need for the different roles involved in this distributed system to understand each other's roles and to build trust between the different roles. This 'interpersonal glue' of trust and access to expertise is critical for optimal and safe care, and needs to be considered when designing interventions for improving out-of-hours palliative care. This study shows that going from analysis to solution(s) is not an easy process. We need to think about it as a learning cycle with multiple steps where ideas are debated and confronted, data and experiences are shared and sensemaking is developed. This calls for a collaborative process with genuine participation and involvement of multiple perspectives. The HCD process also needs to foster imagination and creativity.

Coming up with new and innovative solutions to address problems identified during the analysis phase may be constrained by a lack of time, insufficient resources and poorly designed or managed HCD processes. In previous chapters, I described models of HCD (e.g. the ISO model and the Double Diamond Model) and SEIPS case studies that provide insights on how to design and manage effective and efficient HCD approaches. The HCD process itself is embedded in a work system that needs to foster creativity. Different tools and methods can be used to support creativity during HCD, such as brainstorming (Jun et al., 2024). In addition, we need to fight the tendency of people to propose 'additive changes' and overlook 'subtractive changes' (Adams et al., 2021). The distinction between subtractive and additive changes is particularly relevant when addressing health care safety, given the high workload and overburdened patients, care partners and health care workers; we need to avoid adding work and increasing burden. Subtractive and additive changes are 'action categories that remove components from or add components to the original, respectively' (Adams et al., 2021, p. 258). Instead of adding a task, a technology or a step to an already busy and burdened work system (i.e. additive changes), we should come up with solutions where things are removed (i.e. subtractive changes). An example of this approach is the 'Getting Rid of Stupid Stuff' program aimed at reducing the workload associated with the EHR technology, such as burdensome documentation (Ashton, 2018). This program could be adapted to address the workload associated with all work system elements, including the *technology*. This would involve identifying all work system elements that add to workload, such as inefficient *tasks* and poorly designed *technologies* and *physical environments*. As stated by Coiera (2011), 'In a system that is overconstrained with competing demands, the human attention and physical resources needed to make a new intervention succeed are just not available' (page 3). Addressing high workload in health care is a major endeavor that would benefit from SEIPS and, if successful, would create room for creativity and innovation to improve health care safety.

6.1.2 Integration of SEIPS and HFE Knowledge in HCD

Design is challenging as the HFE knowledge that needs to be integrated into the HCD process is vast. As discussed in previous chapters, the HFE knowledge is broad and deep and continues to grow. Some HFE knowledge may be hard to find, understand or apply. Some HFE design principles may be vague and challenging to apply, especially as their relevance and application are likely to depend on the specific work system and care process under study. We need HFE in HFE; that is, HFE concepts, approaches and methods need to be applied to HFE and HCD. Is the knowledge that we (HFE professionals) produce usable and useful? How does this knowledge fit into the HCD work system and process?

HFE knowledge is often instantiated in guidelines and guidance documents, such as HFE handbooks. Research shows that many of these HFE resources lack usability (Burns et al., 1997; Choi et al., 2006) and do not match the mental models and workflows of designers (Vicente et al., 1998). Burns et al. (1997) made four recommendations for improving HFE guidance documents:

1. detailed application-specific information
2. match between the document and the needs of the designers
3. information to designers about where to find HFE guidance documents
4. organizational policies that foster designers' search for HFE information.

Some effort has been made to improve access to HFE knowledge. For instance, in line with some of the recommendations of Burns et al. (1997), the 'Handbook of Human Factors in Medical Device Design' (Weinger et al., 2011) includes specific guidelines in each of its 19 chapters (e.g. environment of use, anthropometry and biomechanics). Medical device designers can then translate the guidelines into product requirements. Some HFE design principles remain general and cannot be written as specific guidelines. As described in Chapter 3, the Balance Theory has two core principles for work system design: that is, eliminating work system barriers and balancing the work system (Carayon, 2009; Smith & Carayon-Sainfort, 1989). These HFE design principles can be used for guiding the discussion on potential solutions in a workshop or design session, where the results of a SEIPS-based analysis of work system barriers and facilitators are presented. Embedding HFE professionals in health care is critical to support the effective and appropriate use of the vast HFE knowledge for improving health care safety.

The challenge of integrating HFE knowledge has been examined in engineering design processes. Several studies have identified challenges and barriers to the integration of HFE in engineering design projects (Perrow, 1983), such as offshore platform design (Wulff et al., 1999a, 1999b) and the design of accessible medical equipment (Carayon, Grenier, and Alvarado, 2007c). Barriers include the *organizational environment* of engineering design that emphasizes efficiency and speed, and limits access to resources such as HFE professionals and experts. This research points to the need for understanding the work and work system in engineering design projects (Carayon, Grenier, and Alvarado, 2007c). This recommendation also applies to health care work system design projects. We need to understand the work system in which HCD for health care safety occurs:

- Who are the *people* participating in the HCD process?
- What are the *tasks* involved in the HCD process?
- What *tools and technologies* are used in the HCD process?
- *Where* does HCD take place?
- How is the HCD process *organized*?

Understanding the work system of HCD and its barriers and facilitators is an important area of research. We made a similar recommendation to improve the usability of health IT; we need to understand the work and work system involved in designing technologies (Carayon & Hoonakker, 2019). We know little about the work of health IT designers and implementers. Without this knowledge, we do not understand barriers in their work system, in particular those related to the integration of HFE knowledge, such as heuristics to improve the usability of health IT.

6.1.3 Who Are the Designers?

We are all patients. We interact with health care as patients and/or care partners. We are also all designers.

There is evidence of work system design by health care workers. Different terms have been used to describe this process, such as job crafting (Demerouti, 2014;

Wrzesniewski & Dutton, 2001) and tailoring (Mørch & Mehandjiev, 2000). Health care workers adjust and adapt their work system to make it fit better, in particular, the design of their tools and technologies. Patient status boards are an interesting example of a tool that has developed over time to support coordination within and across health care teams. Wears et al. (2006) compared the design and use of patient status boards in an ED and a pediatric hospital unit. They showed how patient status boards were designed over time by health care workers to support coordination of teams within the ED and the pediatric inpatient unit, as well as coordination with other teams, units and health care organizations. For instance, the ED status board uses color to indicate team assignment and various objects (e.g. magnets) for other purposes (e.g. specific hazards or conditions). Watson et al. (2004) observed tailoring practices of anesthesiologists in adapting auditory alarms on their monitoring equipment, such as adjusting the lower limit of the alarm so that a reminder would be sent to the anesthesiologist busy with other demanding tasks.

We also have some evidence that health care workers and patients tailor their work systems to improve their collaborative work. Obradovich and Woods (1996) studied the in-home use of an infusion pump by pregnant women experiencing pre-term labor. First, a nurse sets up the infusion pump for the administration of terbutaline medication for each patient. Then, the nurse trains the patient on tasks she needs to perform for safe therapy at home. Over time, nurses and patients develop various tailoring strategies 'to work around error-prone tasks and device deficiencies' (page 579). These strategies include: development of a patient guide and modification of procedures used by the patient. Health care workers and patients are designers of their own work systems (Obradovich & Woods, 1996).

Eason (2014) describes the adaptive and learning capabilities of people as they 'are not just resources in a work system but are also designers of work systems' (page 215). Therefore, it is important to take advantage of the knowledge and skills of people at the 'sharp end' (e.g. physicians, nurses, patients) as well as professionals with HFE knowledge and skills during HCD projects; this is the essence of collaborative design or co-design. The case studies described in Chapter 5 show the involvement of different perspectives, roles and disciplines in the co-design of work system interventions. All the people involved in the projects were designers of the interventions, including the HFE professionals who managed and facilitated the projects. HFE is design driven (Dul et al., 2012); therefore, HFE professionals should be active participants in HCD and are designers themselves.

6.2 CONTINUOUS DESIGN AND LEARNING

In this book, I propose a broad approach to health care safety that includes both patient safety and worker safety, and their multiple dimensions; see Chapter 1 and Tables 1.1 and 1.2. Improving health care safety, therefore, requires a cycle of analysis and design of solution(s) that produce 'systems changes' (Wood & Wiegmann, 2020). As discussed in Section 6.1.3, it is not easy to address specific health care safety problems and to come up with solutions based on SEIPS analysis. In addition, we need to think about linkages between SEIPS-based design and implementation of interventions; this calls for an approach based on 'muddling through' where design and implementation are linked and integrated. It is important to dedicate energy, skills and resources to the design and implementation of work system interventions for health care safety, but this is not sufficient. We need sustained long-term commitment and leadership for health care safety improvement. This implies support for a continuous design and learning process.

6.2.1 Design for Incompletion and Muddling Through

Given the dynamic nature of health care work systems and a rapidly changing external environment, it is likely that, as soon as a design solution has been found and an intervention is implemented, the work system and the care process have already changed. In addition to being difficult to design, the solution may no longer fit or be sufficient at the time of implementation. The STS Theory provides principles that could help to deal with this dynamic situation; see the principles in Table 2.4 in Chapter 2. One approach is to 'design for incompletion' (Cherns, 1976). This sociotechnical design principle implies that, when an intervention is implemented, people in the work system have some discretion to develop strategies to cope with new circumstances and emergent needs (Eason, 2014). This principle is related to another sociotechnical design principle, that is, minimal criteria specification (Cherns, 1976, 1987). When specifying a work system redesign, we should come up with minimum specifications; this leaves flexibility at the local work system level. These sociotechnical design principles foster autonomy and job control at the local work system level, which can enhance satisfaction and motivation of the people at the center of the work system. Autonomy and job control allow people at the sharp end to develop strategies to deal with unexpected disturbances.

We need to support the creative, adaptive and learning capabilities of people 'so that they can invent solutions in real time to problems that could not have been anticipated in advance' (Flach, 2012, p. 196). Therefore, design needs to be more closely connected with implementation. Designing an intervention for improving health care safety needs to consider the work system and care process in which the intervention will be implemented; therefore, design and implementation of the intervention are interrelated. Muddling through in the interrelated design and implementation processes is a critical mechanism to cope with complexity (Flach, 2012; Norman & Stappers, 2015), such as the complexity found in health care work systems and processes. The political scientist, Charles Lindblom, coined the term 'muddling through' to describe the incremental process of policymaking aimed at addressing complex problems (Lindblom, 1959). In designing and implementing work system redesigns to improve health care safety, muddling through involves trials and errors, pilot studies and small and incremental adjustments to work system redesign (Flach, 2012; Norman & Stappers, 2015). This is best accomplished in collaborative design projects where everyone involved needs to deal with ambiguity and uncertainty, and accept compromises (Détienne, 2006; Norman & Stappers, 2015).

6.2.2 Continuous Design and Learning Versus Implementation

There seems to be an artificial boundary between design and implementation. But design and implementation are interrelated. Implementation is not something that is done after the design is finished. Thinking about the implementation of the work system intervention should start when design solutions are discussed. This leads me to propose the concept of continuous design.

Implementation science has emerged as a major movement in health care quality and patient safety (Nilsen, 2024; Wilson & Kislov, 2022). There is a journal dedicated to 'Implementation Science'. Funding agencies have dedicated resources to research projects aimed at developing and using implementation science with the goal of improving the implementation of 'evidence-based health care'. Criticisms of implementation science include the proliferation of models and frameworks (Wilson & Kislov, 2022). In addition, the utility of these models and frameworks may be questionable, in particular in light of pragmatic approaches such as HCD (Wilson & Kislov, 2022). When trying to improve health care safety, implementation science has not paid insufficient attention to the 'what', that is, the intervention that is implemented. In health care safety, early efforts have focused on identifying 'evidence-based practices'; see, for instance, AHRQ reports on patient safety practices (Shekelle, Wachter, et al., 2013; Shojania et al., 2001). However, we know that this is not sufficient. We need to address the design of work systems and care processes in a deeper manner; this is what I propose in this book, with a focus on SEIPS. We need to pay attention to the design of the work system intervention and its fit in the ever-changing work system and care process.

I do not negate the critical role of implementation in supporting the use of work system interventions. In Chapter 4, I describe organizational principles that should be applied when implementing a work system intervention. However, paying attention to the implementation of a specific intervention is not sufficient; we need to achieve systems changes at all system levels (Wood & Wiegmann, 2020). As emergence is a key principle in systems thinking and for SEIPS, we should expect 'surprises' when an intervention is implemented; therefore, our effort to improve health care safety does not stop after the implementation of an intervention. We need to move toward a process of continuous design (Carayon, 2022; Carayon & Salwei, 2021; Falzon, 2007). As stated by Norman and Stappers (2015), 'The design process never ends' (page 91). In Section 6.1.3 of this chapter, I describe the role of health care workers and patients as designers when they tailor and adapt their work systems. This supports the idea that design is continuous and does not end after the implementation of work system interventions.

Weick and Quinn (1999) contrast episodic change with continuous change. Episodic changes tend to be infrequent, have a broad scope and be disruptive, whereas continuous changes are ongoing, evolving and cumulative. In episodic changes, the emphasis is on planning for the change and paying attention to implementation strategies to anticipate resistance to change and ensure acceptance of the change. In continuous changes, the focus is on learning of individuals, teams and organizations.

In a study of the implementation of smart infusion pump technology in a hospital, we observed an episodic change followed by a continuous change (Carayon et al., 2008). See details of this case study in the following box.

BOX – EPISODIC AND CONTINUOUS CHANGES IN THE IMPLEMENTATION OF SMART INFUSION PUMP TECHNOLOGY IN A HOSPITAL

Before the new infusion pump technology was implemented, an interdisciplinary infusion pump team (including HFE professionals) analyzed the new technology and its fit in the medication administration process using the Failure Mode and Effects Analysis (FMEA) method (Wetterneck et al., 2006). The potential hazards, their contributing factors and potential solutions were analyzed using the work system model. A pilot test of the new infusion pump was conducted in two hospital units. These pre-implementation activities helped to improve the design of the drug library and alarms of the pump (*technology element of the work system*), and to refine the training materials and process for the future pump users, that is, nurses and anesthesiologists (*task and organization elements of the work system*).

On October 21, 2003, the new infusion pump technology was implemented throughout the hospital. Three weeks later, a major adverse event occurred during a surgery (Schroeder et al., 2006): a tubing misload led to free flow of an anesthetic. The event was discovered immediately. The patient suffered only momentary consequences and fully recovered. The pump was immediately sequestered, and an investigation was launched with support from the HFE members of the interdisciplinary infusion pump team. An external company was hired to conduct an additional evaluation of the event. Multiple meetings took place between the hospital and the pump manufacturer. Several redesigns of the pump were proposed, and some went through usability testing and had to be further refined (Hundt et al., 2005). Seven months later, a newly redesigned pump was implemented.

This case study shows characteristics of both an episodic change (planning activities before the implementation of the infusion pump) and a continuous change (activities after the adverse event, including organizational learning and redesigns of the infusion pump after usability testing). The episodic change and the continuous change were related. Some of the same people who participated in the pre-implementation activities, such as the FMEA, also participated in activities after the adverse event. A collaborative process was established pre-implementation; the collaboration was key to building trust and respect and to creating the learning environment that was so important after the adverse event. This case study had many of the 'good' principles to manage the episodic change related to the implementation of the infusion pump. However, an adverse event occurred; this was an unanticipated consequence that was quickly identified and led to collaborative learning during the continuous change process.

We need to pay attention to the interactions between design and implementation, closely monitor and evaluate what happens after the implementation of a health care safety intervention and learn from the implementation to continually improve the work system and care process. This is what I mean by continuous design; it is an organizational learning process, which fits the feedback loops in SEIPS (Carayon & Salwei, 2021).

6.2.3 Continuous Design and Learning Versus Sustainability

Sustainability is often discussed in the health care safety literature and refers to multiple 'objects'. Let us look at different examples of what is sustained. In the Keystone Michigan Project, Pronovost et al. (2010) were concerned about sustaining improvement in HAI reduction, that is, sustaining a safety outcome. Moon et al. (2022) related sustaining outcome improvement to sustaining quality and safety improvement programs within hospitals. They found that multiple organizational factors, for example, leadership, organizational culture and coordination and organization of changes, contributed to sustaining quality and safety improvement initiatives in hospitals. At a national level, Parand et al. (2012) identified factors that helped to sustain a patient safety improvement collaborative, such as organizational capability and integration with existing structures and processes. A review of various quality and safety improvement programs mentions sustaining change, enthusiasm, commitment, effort, collective knowledge and momentum (Dixon-Woods et al., 2012). Overall, in the health care safety improvement literature, sustainability refers to different objects and components of SEIPS:

- sustaining changes in the design of the work system and care process (in particular, the organizational element)
- sustaining outcome improvement
- sustaining feedback and organizational change.

In this book, I argue that SEIPS and HFE are critical to health care safety improvement. Therefore, to ensure that positive changes are sustained and that we learn from failures, I suggest that we also need to sustain the integration of HFE skills and knowledge throughout health care.

To address the complexity of health care systems and care processes and improve health care safety over time, we need to build learning and sharing mechanisms and an organizational culture, where systems approaches, such as SEIPS and HFE, are valued, recognized and applied. This cannot be done for just one project; it needs to be embedded in the fabric of the organization. Creating and sustaining a positive organizational cultural change is critical to supporting the integration of SEIPS and HFE. A review of the literature on sustaining organizational cultural change in health care provides ideas for how to accomplish this (Willis et al., 2016). For instance, building on an organization's sense of experimentation, SEIPS could be tried in incremental change projects; this would build knowledge and confidence in SEIPS and, more broadly, HFE. This incremental approach fits with the concept of muddling through proposed by Lindblom (1959) and, over time, could show the benefits of an HFE-based systems approach such as SEIPS.

6.3 WHOSE WORK ARE WE (RE)DESIGNING?

In HFE and SEIPS, the person is at the center of the work system. That person could be an individual, e.g. a physician, nurse, patient or care partner, or it could be a group of individuals or a care team. HFE has been traditionally applied to formal work and jobs, such as those found in health care (e.g. pharmacist, nurse, therapist, physician). In health care, the application of HFE has been expanded to other types of work, such as work done by patients in managing their medication or treatment and work done by care partners in supporting and helping their loved ones. Initially, SEIPS applications focused on health care workers. As the SEIPS family developed, we expanded the person at the center of the work system and included the patient and care partner. Could we go one step further and include the community at large? I explore this question in the section on community ergonomics.

6.3.1 Patient-Centered HFE

In SEIPS, the role of the patient has been described in multiple different ways. See Table 3.1 in Chapter 3 for an overview of the role of patients in different SEIPS models. In the original SEIPS model (Carayon, Hundt, Karsh, et al., 2006) and its updated version of 2014 (Carayon, Wetterneck, Rivera-Rodriguez, et al., 2014), the patient was involved in the work system of health care workers; for instance, a nurse administering a medication to a patient [*task element of the nurse's work system*]. The patient was also mentioned as being at the center of the work system. The SEIPS 2.0 model clarified the person at the center of the work system and described three types of work: work done by a healthcare professional, patient work and collaborative work (Holden et al., 2013). The SEIPS 3.0 model focuses on the patient journey and the patient's interactions with multiple health care organizations and care settings (Carayon, Wooldridge, et al., 2020). The SEIPS 101 describes people very broadly, including both health care workers and patients (Holden & Carayon, 2021).

The SEIPS family has been expanded to deepen our understanding of the work system of patients. The Patient Work System model defines the work system as interactions between *people (i.e. patient, health care professional and care partner), tasks, technologies, physical-spatial context, social-cultural context and organizational context* (Holden, Schubert, and Mickelson, 2015). The Patient Work System model was used in an analysis of barriers experienced by patients with heart failure and their care partners (Holden, Schubert, and Mickelson, 2015) and in exploring how adolescents with type 1 diabetes manage their health alongside their parents

(Valdez et al., 2024). The consolidated model of the macroergonomic patient work system was developed by an interdisciplinary group of researchers involved in three separate projects looking at the work of patients with chronic illnesses (Holden et al., 2017). The model focuses on the home- and community-based work systems of patients. It includes the microergonomic interacting elements of *people, task and technology* and three macroergonomic factors, i.e. *physical, social and organizational context*. The Patient Work System and the consolidated model of the macroergonomic patient work system have many similarities to the SEIPS family described in Table 3.1

Over the years, research on patient-centered HFE has increased (Holden et al., 2020; Werner, Ponnala, et al., 2020). A review of the literature published between 2007 and 2017 on patient-centered HFE or patient ergonomics shows that most research focused on older patients or patients with chronic illnesses. Recently, patient-centered HFE research has addressed issues specific to children and adolescents (Barton et al., 2021; Valdez et al., 2024) and the work of care partners (Scheer et al., 2024; Werner et al., 2022). Holden and Valdez edited a series of two books on patient ergonomics (Holden & Valdez, 2021; Valdez & Holden, 2021). In addition to understanding the work of patients, development of the SEIPS family includes a specific focus on the work of care partners (Holden, Schubert, and Mickelson, 2015; Werner, Ponnala, et al., 2020). Since health care safety is also safety for care partners (see Chapter 1), we need to apply SEIPS to the work system of care partners with the goal of improving safety for both patients and care partners.

The SEIPS family could further expand patient-centered HFE to address the linkages between health care and health. The SEIPS 3.0 model describes the patient journey and emphasizes health care safety at the interfaces of various health care organizations and care settings (Carayon, Wooldridge, et al., 2020). What happens in between episodes, where the person is not 'just a patient'? We need to look at the patient journey across the life span, not just one sequence or one episode of care. We need to look at what happens between the different sequences of the patient journey, i.e. the health journey. The patient journey focuses on encounters with health care, whereas the health journey 'includes a lifetime of other encounters and experiences that can enhance responsible healthy behaviors' (Jones, 2013, p. 2). SEIPS has been used to analyze and improve care processes and could be expanded to improve health-seeking processes during the health journey. SEIPS could also be applied to the domain of public health to understand the barriers and facilitators in multiple systems and levels to improve, for instance, vaccination and preventive care. This calls for the application of SEIPS to examine broad safety issues in local, national and international communities.

6.3.2 Community Ergonomics

Over the years, several HFE experts have called for the HFE discipline to tackle broad societal issues. In the domain of IT, Bjorn-Andersen (1988) described three different levels of HFE application: workplace ergonomics (e.g. design of keyboard and computer interfaces), organizational ergonomics (e.g. designing IT while simultaneously enhancing social contact and improving job content) and societal ergonomics. Societal ergonomics focuses on unemployment and the role of technology in automating jobs and possibly contributing to unemployment. Bjorn-Andersen (1988) called for attention to all three levels of HFE application for the attainment of human values and making HFE 'human'. Moray (2000) encouraged HFE professionals to address major societal problems, such as 'the shortage of water, pollution, urbanization and overcrowding, an ageing population, and climate change' (page 859). Around the world, HFE professionals and researchers have begun to tackle major societal problems, such as climate change and environmental sustainability (Thatcher & Yeow, 2016) and security and safety of energy supplies (Meshkati et al., 2016). Community Ergonomics is an approach that goes beyond macroergonomics or organizational ergonomics and aims at addressing societal issues of community health and well-being (Smith et al., 2002).

The Community Ergonomics approach was initiated at the University of Wisconsin-Madison in the United States in the late 1990s (Smith et al., 2002; Smith & Smith, 1994), and initially focused on communities and neighborhoods characterized by poverty, social isolation and dependency (Cohen & Smith, 1994; Smith et al., 1994). Over the years, the Community Ergonomics approach has been applied to the design of corporate social responsibility initiatives to balance economics with social awareness (Smith et al., 2009), and the design of a network of enterprises in a French region to improve quality in the enterprises as well as quality of life and work in the region (Smith et al., 2002). Barton et al. (2025) described six Community Ergonomics projects in a variety of community contexts (e.g. low-income racial and ethnic marginalized communities, rural caregivers of people living with dementia in the United States and South African smallholder crop farmers) and countries (i.e. United States, South Africa, Colombia). The Community Ergonomics approach builds on the STS Theory and proposes a conceptual systems model with the following components (Smith et al., 2002):

- community social sub-system, i.e. community residents
- community technical sub-system (e.g. institutions, services, policies)
- community ergonomics process, i.e. linking the social and economic sub-systems and the external environment.

Smith et al. (2002) defined seven principles of Community Ergonomics, including action-oriented, participatory and integrating multiple perspectives; see Table 6.1 for the complete list of Community Ergonomics principles. Community Ergonomics is closely related to SEIPS and HCD, as shown in Table 6.1. Future development of SEIPS could further integrate the Community Ergonomics principles and tackle major health care safety problems at national and international levels. This fits with the humanity-centered design approach described by Don Norman in his 2023 book on 'Design for a Better World'. Don Norman (2023) proposes to move from human-centered design to humanity-centered design to address the societal problems of sustainability, inequity and bias.

TABLE 6.1
Principles of Community Ergonomics and their Linkage to SEIPS and HCD

Principles of Community Ergonomics	Description	Linkage to SEIPS and HCD
Action-oriented approach	Addressing issues of importance to the community. Focus on changing the system characteristics and conditions.	Improving health care safety, i.e. a combination of patient safety and worker safety. Focus on changing the work system and care process.
Participation by everyone	Participation by community members and stakeholders. Design with and by the community.	Participation in all stages of HCD. Design with and by users.
Diversity and conflict management	Process for managing diversity, conflict and confusion. Coordination, communication and integration of diverse information and perspectives.	Multiple perspectives represented in HCD. Collaboration, development of common ground and negotiation in HCD.
Encouraging learning	Learning by community members, e.g. transfer of knowledge and skills.	Individual and organizational learning as a feedback loop in SEIPS.
Building self-regulation	Providing community members with opportunities to increase their ability and capacity for self-regulation.	Collaborative design and genuine participation.
Feedback triad	Reactive feedback, e.g. perception of impact. Instrumental feedback, e.g. goals achieved, milestones reached. Operational feedback, e.g. new policies, laws or institutions.	Feedback loops in SEIPS. SEIPS evaluation of work system interventions with data on the impact of intervention on work system, care process and outcomes.
Continuous improvement and innovation	Need for ongoing evaluation and monitoring of the community's needs, desires and values.	Feedback loops in SEIPS. Adaptation, improvement and learning. Continuous design.

7 Conclusion

We are all patients; we are all designers. We are all together.

Improving health care safety is about making changes for and with people. It is a moral and ethical endeavor that is based on humanistic values, such as respect, empathy and humility. The HFE professionals involved in SEIPS and HCD, therefore, need to demonstrate skills and abilities that support these values. They need to be good listeners and build trust with the people, organizations and communities they work and interact with. They need to show empathy and be humble as they deal with potentially challenging human consequences and sensitive problems of health care safety. As HCD involves multiple perspectives, they need to accept ambiguity and uncertainty and be able to deal with and manage tension and conflicts.

Improving health care safety means working together. Over the years, I had the chance to work with multiple health care organizations and various collaborators. This collaborative work to improve health care safety requires trust, respect, communication and dedication. Transparent and open mechanisms help to share information and create opportunities for learning for everyone involved. It is also important to get involved at various system levels where HFE and systems approaches such as SEIPS need to be recognized, supported and integrated; this involves genuine collaboration with the broad health care sector and national and international policies and programs.

In this book, I propose SEIPS as a model, a conceptual framework, a way of thinking in systems, and a toolbox of methods and processes. SEIPS is a piece of the puzzle in our collective efforts at improving health care safety around the world, but it is only one piece of the puzzle. Many other efforts, disciplines and approaches are necessary. Therefore, I would like to encourage everyone involved in health care safety to be curious and open and collaborate with others to enrich your knowledge and experience. We can learn a lot from each other; this is important and critical for building a safer and healthier world.

DOI: 10.1201/9781003480310-7

References

Abebe, E., & Chui, M. A. (2018). Exploring barriers to medication safety in an Ethiopian hospital emergency department: A human factors engineering approach. *Innovations in Pharmacy*, *9*(1), 1–10.

Abraham, O., LeMay, S., Nixon, G., Braun, A., Decker, C., & Szela, L. (2021). Characterizing medication management and the role of pharmacists in caring for people living with cystic fibrosis: A work system approach. *Journal of the American Pharmacists Association*, *61*(5), 547–554.

Acher, A. W., LeCaire, T. J., Hundt, A. S., Greenberg, C. C., Carayon, P., Kind, A. J., & Weber, S. M. (2015). Using human factors and systems engineering to evaluate readmission after complex surgery. *Journal of the American College of Surgeons*, *221*(4), 810–820.

Ackoff, R. L. (1971). Towards a system of systems concept. *Management Science*, *17* (11), 661–671.

Adam, T., & de Savigny, D. (2012). Systems thinking for strengthening health systems in LMICs: Need for a paradigm shift. *Health Policy and Planning*, *27*(Suppl 4), 1–3.

Adams, G. S., Converse, B. A., Hales, A. H., & Klotz, L. E. (2021). People systematically overlook subtractive changes. *Nature*, *592*(7853), 258–261.

Adelman, R. D., Tmanova, L. L., Delgado, D., Dion, S., & Lachs, M. S. (2014). Caregiver burden: A clinical review. *JAMA*, *311*(10), 1052–1060.

Agency for Healthcare Research and Quality (AHRQ). (2024). *National Healthcare Quality and Disparities Report Chartbook on Patient Safety*. Agency for Healthcare Research and Quality.

AHRQ PSNet. (2024). *What is Patient Safety?* Retrieved October 23, 2024 from https://psnet.ahrq.gov/patient-safety-101

Aiken, L. H., Lasater, K. B., Sloane, D. M., Pogue, C. A., Fitzpatrick Rosenbaum, K. E., Muir, K. J., McHugh, M. D., & US Clinician Wellbeing Study Consortium. (2023). Physician and nurse well-being and preferred interventions to address burnout in hospital practice: Factors associated with turnover, outcomes, and patient safety. *JAMA Health Forum*, *4*(7), e231809.

Al Juffali, L., Al-Aqeel, S., Knapp, P., Mearns, K., Family, H., & Watson, M. (2019). Using the Human Factors Framework to understand the origins of medication safety problems in community pharmacy in Saudi Arabia: A qualitative study. *Research in Social and Administrative Pharmacy*, *15*(5), 558–567.

Alfred, M., Catchpole, K., Huffer, E., Fredendall, L., & Taaffe, K. M. (2020). Work systems analysis of sterile processing: Decontamination. *BMJ Quality & Safety*, *29*(4), 320–328.

Alfred, M., Catchpole, K., Huffer, E., Fredendall, L., & Taaffe, K. M. (2021). Work systems analysis of sterile processing: Assembly. *BMJ Quality & Safety*, *30*(4), 271–282.

Alli, B. O. (2008). *Fundamental Principles of Occupational Safety and Health*. International Labour Organization.

Allread, W. G., & Israelski, E. W. (2010). Anthropometry and biomechanics. In M. B. Weinger, M. E. Wiklund, & D. J. Gardner-Bonneau (Eds.), *Handbook of Human Factors in Medical Device Design* (pp. 97–152). CRC Press.

Alvarado, C. J. (2012). The physical environment in health care. In P. Carayon (Ed.), *Handbook of Human Factors and Ergonomics in Health Care and Patient Safety* (Second ed., pp. 215–234). Taylor & Francis Group.

Andersen, L. L., Vinstrup, J., Villadsen, E., Jay, K., & Jakobsen, M. D. (2019). Physical and psychosocial work environmental risk factors for back injury among healthcare workers: Prospective cohort study. *International Journal of Environmental Research and Public Health*, *16*(22), 1–10.

Anderson, J. E., Ross, A. J., Back, J., Duncan, M., Snell, P., Walsh, K., & Jaye, P. (2016). Implementing resilience engineering for healthcare quality improvement using the CARE model: A feasibility study protocol. *Pilot and Feasibility Studies*, *2*, 1–9.

Antonacci, G., Lennox, L., Barlow, J., Evans, L., & Reed, J. (2021). Process mapping in healthcare: A systematic review. *BMC Health Services Research*, *21*(1), 1–15.

Arbaje, A. I., Kansagara, D. L., Salanitro, A. H., Englander, H. L., Kripalani, S., Jencks, S. F., & Lindquist, L. A. (2014). Regardless of age: Incorporating principles from geriatric medicine to improve care transitions for patients with complex needs. *Journal of General Internal Medicine*, *29*(6), 932–939.

Arcuri, R., Bulhões, B., Jatobá, A., Bellas, H. C., Koster, I., d'Avila, A. L., Vidal, M. C. R., Burns, C. M., & De Carvalho, P. V. R. (2020). Gatekeeper family doctors operating a decentralized referral prioritization system: Uncovering improvements in system resilience through a grounded-based approach. *Safety Science*, *121*, 177–190.

Ash, J. S., Sittig, D. F., Dykstra, R., Campbell, E., & Guappone, K. (2009). The unintended consequences of computerized provider order entry: Findings from a mixed methods exploration. *International Journal of Medical Informatics*, *78*(Supplement 1), S69–S76.

Ashton, M. (2018). Getting rid of stupid stuff. *New England Journal of Medicine*, *379*(19), 1789–1791.

Atkinson, M. K., Benneyan, J. C., Bambury, E. A., Schiff, G. D., Phillips, R. S., Hunt, L. S., Belleny, D., & Singer, S. J. (2022). Evaluating a patient safety learning laboratory to create an interdisciplinary ecosystem for health care innovation. *Health Care Management Review*, *47*(3), E50–E61.

Augustsson, H., Churruca, K., & Braithwaite, J. (2020). Change and improvement 50 years in the making: A scoping review of the use of soft systems methodology in healthcare. *BMC Health Services Research*, *20*(1), 1063.

Baartmans, M. C., Van Schoten, S. M., Smit, B. J., & Wagner, C. (2023). Using the generic analysis method to analyze sentinel event reports across hospitals: A retrospective cross-sectional study. *Journal of Patient Safety*, *19*(3), 158–165.

Baartmans, M. C., Van Schoten, S. M., & Wagner, C. (2022). Generic analysis method to learn from serious adverse events in Dutch hospitals: A human factors perspective. *BMJ Open Quality*, *11*(1), 1–9.

Bainbridge, L. (1983). Ironies of automation. *Automatica*, *19*(6), 775–779.

Bakker, A. B., & Demerouti, E. (2017). Job demands-resources theory: Taking stock and looking forward. *Journal of Occupational Health Psychology*, *22*(3), 273–285.

Balogh, E. P., Miller, B. T., & Ball, J. R. (2015). *Improving Diagnosis in Health Care*. The National Academies Press.

Bamgboje-Ayodele, A., Boscolo, A., Newton, N., Tariq, A., Dao, M., Shaw, M., Hutchings, O., Shaw, T., McPhail, S. M., & Baysari, M. (2025). Detection and management of clinical deterioration in a virtual hospital: A work system analysis. *Applied Ergonomics*, *127*, 104530.

Barcellini, F., Cerf, M., & Lacomblez, M. (2024). Developmental foundations of Activity-Centered Ergonomics: Knowledge encounters to construct both a critical analysis of work and developmental set-ups. *Ergonomics*, *68*(6), 813–831.

Barker, A. K., Brown, K., Siraj, D., Ahsan, M., Sengupta, S., & Safdar, N. (2017). Barriers and facilitators to infection control at a hospital in northern India: A qualitative study. *Antimicrobial Resistance & Infection Control*, *6*(1), 35.

Barker, A. K., Ngam, C., Musuuza, J. S., Vaughn, V. M., & Safdar, N. (2017). Reducing Clostridium difficile in the inpatient setting: A systematic review of the adherence to and effectiveness of C. difficile prevention bundles. *Infection Control & Hospital Epidemiology*, *38*(6), 639–650.

Barker, L. M., & Nussbaum, M. A. (2011). Fatigue, performance and the work environment: A survey of registered nurses. *Journal of Advanced Nursing*, *67*(6), 1370–1382.

Barton, H. J., Coller, R. J., Loganathar, S., Singhe, N., Ehlenbach, M. L., Katz, B., Warner, G., Kelly, M. M., & Werner, N. E. (2021). Medical device workarounds in providing care for children with medical complexity in the home. *Pediatrics*, *147*(5), 1–10.

Barton, H. J., Harris, M. T., Rogers, C. C., Lange-Morales, K., Thatcher, A., Newman, L. C., Wooldridge, A. R., Rogers, W. A., Montague, E., Werner, N. E., & Valdez, R. S. (2025). Leveraging Ergonomics and Human Factors (E/HF) for community impact: What have we learned about how to make a difference. *Ergonomics*, 1–19.

Barton, H. J., Salwei, M. E., Rutkowski, R. A., Wust, K., Krause, S., Hoonakker, P. L., Dail, P. V., Buckley, D. M., Eastman, A., Ehlenfeldt, B., Patterson, B. W., Shah, M. N., King, B. J., Werner, N. E., & Carayon, P. (2023). Evaluating the usability of an emergency department after visit summary: Staged heuristic evaluation. *JMIR Human Factors*, *10*, e43729.

Batalden, M., Batalden, P., Margolis, P., Seid, M., Armstrong, G., Opipari-Arrigan, L., & Hartung, H. (2016). Coproduction of healthcare service. *BMJ Quality & Safety*, *25*(7), 509–517.

Bates, D. W., Cullen, D. J., Laird, N., Petersen, L. A., Small, S. D., Servi, D., Laffel, G., Sweitzer, B. J., Shea, B. F., Hallisey, R., Vander Vliet, M., Nemeskal, R., & Leape, L. L. (1995). Incidence of adverse drug events and potential adverse drug events: Implications for prevention. *JAMA*, *274*(1), 29–34.

Bates, D. W., Levine, D. M., Salmasian, H., Syrowatka, A., Shahian, D. M., Lipsitz, S., Zebrowski, J. P., Myers, L. C., Logan, M. S., Roy, C. G., Iannaccone, C., Frits, M. L., Volk, L. A., Dulgarian, S., Amato, M. G., Edrees, H. H., Sato, L., Folcarelli, P., Einbinder, J. S., … Mort, E. (2023). The safety of inpatient health care. *New England Journal of Medicine*, *388*(2), 142–153.

Bates, D. W., & Singh, H. (2018). Two decades since To Err Is Human: An assessment of progress and emerging priorities in patient safety. *Health Affairs*, *37*(11), 1736–1743.

Beede, E., Baylor, E., Hersch, F., Iurchenko, A., Wilcox, L., Ruamviboonsuk, P., & Vardoulakis, L. M. (2020). A human-centered evaluation of a deep learning system deployed in clinics for the detection of diabetic retinopathy *Proceedings of the 2020 CHI Conference on Human Factors in Computing Systems, Honolulu, HI, USA*, April 2020.

Benishek, L. E., Kachalia, A., & Daugherty Biddison, L. (2023). Improving clinician well-being and patient safety through human-centered design. *JAMA*, *329*(14), 1149–1150.

Berenholtz, S. M., Pronovost, P., Lipsett, P. A., Hobson, D., Earsing, K., Farley, J. E., Milanovich, S., Garrett-Mayer, E., Winters, B. D., Rubin, H. R., Dorman, T., & Perl, T. M. (2004). Eliminating catheter-related bloodstream infections in the intensive care unit. *Critical Care Medicine*, *32*(10), 2014–2020.

Berman, L., Kavalier, M., Gelana, B., Tesfaw, G., Siraj, D., Shirley, D., & Yilma, D. (2021). Utilizing the SEIPS model to guide hand hygiene interventions at a tertiary hospital in Ethiopia. *PloS one*, *16*(10), e0258662.

Bernal, D., Campos-Serna, J., Tobias, A., Vargas-Prada, S., Benavides, F. G., & Serra, C. (2015). Work-related psychosocial risk factors and musculoskeletal disorders in hospital nurses and nursing aides: A systematic review and meta-analysis. *International Journal of Nursing Studies*, *52*(2), 635–648.

Bertalanffy, L. V. (1950). An outline of general system theory. *The British Journal for the Philosophy of Science*, *1*(2), 134–165.

Bertalanffy, L. V. (1969). *General System Theory: Foundations, Development, Applications*. G. Braziller.

Berwick, D. M., Nolan, T. W., & Whittington, J. (2008). The triple aim: Care, health, and cost. *Health Affairs*, *27*(3), 759–769.

Bethel, C., Rainbow, J. G., & Johnson, K. (2022). A qualitative descriptive study of the COVID-19 pandemic: Impacts on nursing care delivery in the critical care work system. *Applied Ergonomics*, *102*, 103712.

Beuscart-Zephir, M. C., Pelayo, S., Borycki, E., & Kushniruk, A. (2012). Human factors consideration in health IT design and development. In P. Carayon (Ed.), *Handbook of Human Factors and Ergonomics in Health Care and Patient Safety* (Second ed., pp. 649–670). Taylor & Francis Group.

Beyer, H., & Holtzblatt, K. (1999). Contextual design. *Interactions*, *6*(1), 32–42.

Bjorn-Andersen, N. (1988). Are 'human factors' human?. *The Computer Journal*, *31*(5), 386–390.

Blijleven, V., Koelemeijer, K., Wetzels, M., & Jaspers, M. (2017). Workarounds emerging from electronic health record system usage: Consequences for patient safety, effectiveness of care, and efficiency of care. *JMIR Human Factors*, *4*(4), e27.

Boakye-Dankwa, E., Teeple, E., Gore, R., Procare Research Team, & Punnett, L. (2017). Associations among health care workplace safety, resident satisfaction, and quality of care in long-term care facilities. *Journal of Occupational and Environmental Medicine*, *59*(11), 1127–1134.

Bock, A. (2024). To fix burnout, new initiatives go beyond worker resilience and put onus on health systems. *JAMA*, *331*(24), 2067–2069.

Bodenheimer, T., & Sinsky, C. (2014). From Triple to Quadruple Aim: Care of the patient requires care of the provider. *The Annals of Family Medicine*, *12*(6), 573–576.

Bogner, M. S. (Ed.). (1994). *Human Error in Medicine*. Lawrence Erlbaum Associates.

Bogner, M. S. (2007). The artichoke systems approach for identifying the why of error. In P. Carayon (Ed.), *Handbook of Human Factors in Health Care and Patient Safety* (pp. 109–126). Lawrence Erlbaum Associates.

Bohr, P. C., Evanoff, B. A., & Wolf, L. (1997). Implementing participatory ergonomics teams among health care workers. *American Journal of Industrial Medicine*, *32*(3), 190–196.

Bolis, I., & Sznelwar, L. I. (2016). A case study of the implementation of an ergonomics improvement committee in a Brazilian hospital: Challenges and benefits. *Applied Ergonomics*, *53*(Pt A), 181–189.

Bourne, R. S., Jeffries, M., Phipps, D. L., Jennings, J. K., Boxall, E., Wilson, F., March, H., & Ashcroft, D. M. (2023). Understanding medication safety involving patient transfer from intensive care to hospital ward: A qualitative sociotechnical factor study. *BMJ Open*, *13*(5), e066757.

Bowie, P., Baharnah, A. A., Alkutbe, R., Abid, M. M., Almelaifi, A., & Abid, M. H. (2025). Using Human Factors science to improve quality and safety of healthcare. *Global Journal on Quality and Safety in Healthcare*, *8*(2), 93–96.

Bowie, P., & Vosper, H. (2024). The Systems Engineering Initiative for Patient Safety (SEIPS) - A human factors approach to work system analysis. In C. Cox, H. P. N. Hughes, & J. Nicholls (Eds.), *Patient Safety - Emerging Applications of Safety Science* (pp. 19–40). Class Professional Publishing.

Braithwaite, J., Churruca, K., Ellis, L. A., Long, J., Clay-Williams, R., Damen, N., Herkes, J., Pomare, C., & Ludlow, K. (2017). *Complexity Science in Healthcare - Aspirations, Approaches, Applications and Accomplishments - A White Paper*. Australian Institute of Health Innovation at Macquarie University.

Braithwaite, J., Vincent, C., Garcia-Elorrio, E., Imanaka, Y., Nicklin, W., Sodzi-Tettey, S., & Bates, D. W. (2020). Transformational improvement in quality care and health systems: The next decade. *BMC Medicine*, *18*(1), 340.

Braithwaite, J., Wears, R. L., & Hollnagel, E. (2015). Resilient health care: Turning patient safety on its head. *International Journal for Quality in Health Care*, *27*(5), 418–420.

Bratteteig, T., & Wagner, I. (2014). *Disentangling Participation - Power and Decision-Making in Participatory Design*. Springer.

Bratteteig, T., & Wagner, I. (2016). Unpacking the notion of participation in participatory design. *Computer Supported Cooperative Work (CSCW)*, *25*(6), 425–475.

Brenner, S. K., Kaushal, R., Grinspan, Z., Joyce, C., Kim, I., Allard, R. J., Delgado, D., & Abramson, E. L. (2016). Effects of health information technology on patient outcomes: A systematic review. *Journal of the American Medical Informatics Association*, *23*(5), 1016–1036.

Brown, K. Q., & Mitchell, T. R. (1991). A comparison of Just-In-Time and batch manufacturing: The role of performance obstacles. *Academy of Management Journal*, *34*(4), 906–917.

Brown, T. (2009). *Change by Design: How Design Thinking Transforms Organizations and Inspires Innovation*. Harper Business.

Brownhill, R., & Bowie, P. (2024). Walk-through-talk-through analysis to support healthcare safety and improvement activity. In C. Cox, H. Hughes, & J. Nicholls (Eds.), *Patient Safety: Emerging Applications of Safety Science* (pp. 95–112). Class Professional Publishing.

Burns, C. M., Vicente, K. J., Christoffersen, K., & Pawlak, W. S. (1997). Toward viable, useful and usable human factors design guidance. *Applied Ergonomics*, *28*(5/6), 311–322.

Businger, A. C., Fuller, T. E., Schnipper, J. L., Rossetti, S. C., Schnock, K. O., Rozenblum, R., Dalal, A. K., Benneyan, J., Bates, D. W., & Dykes, P. C. (2019). Lessons learned implementing a complex and innovative patient safety learning laboratory project in a large academic medical center. *Journal of the American Medical Informatics Association*, *27*(2), 301–307.

Cabrera, D., & Cabrera, L. (2019). What is systems thinking? In M. J. Spector, B. B. Lockee, & M. D. Childress (Eds.), *Learning, Design, and Technology: An International Compendium of Theory, Research, Practice, and Policy* (pp. 1–28). Springer International Publishing.

Cairns, C., & Kang, K. (2021, October 11, 2023). *National Hospital Ambulatory Medical Care Survey: 2021 emergency department summary tables*. Retrieved January 10, 2024 from https://ftp.cdc.gov/pub/Health_Statistics/NCHS/Dataset_Documentation/NHAMCS/doc21-ed-508.pdf

Campion, M. A., & Thayer, P. W. (1985). Development and field evaluation of an interdisciplinary measure of job design. *Journal of Applied Psychology*, *70*(1), 29–43.

Campion, M. A., & Thayer, P. W. (1987). Job design: Approaches, outcomes, and trade-offs. *Organizational Dynamics*, *15*(3), 66–79.

Carayon, P. (2006). Human factors of complex sociotechnical systems. *Applied Ergonomics*, *37*(4), 525–535.

Carayon, P. (Ed.). (2007). *Handbook of Human Factors in Health Care and Patient Safety*. Lawrence Erlbaum Associates.

Carayon, P. (2009). The Balance Theory and the work system model… Twenty years later. *International Journal of Human-Computer Interaction*, *25*(5), 313–327.

Carayon, P. (2010). Human factors in patient safety as an innovation. *Applied Ergonomics*, *41*(5), 657–665.

Carayon, P. (Ed.). (2012). *Handbook of Human Factors and Ergonomics in Health Care and Patient Safety*. Taylor & Francis Group.

Carayon, P. (2022). Implementation or continuous design? The contribution of human factors engineering to health care quality and patient safety. In F. Rapport, R. Clay-Williams, & J. Braithwaite (Eds.), *Key Concepts in Implementation Science: Translation and Improvement in Medicine and Healthcare* (pp. 89–92). Routledge

Carayon, P., Albayrak, A., Goossens, R., Hoonakker, P., Hose, B.-Z., Kelly, M., Melles, M., & Salwei, M. (2021). Macroergonomics of patient work - Engaging patients in improving sociotechnical context of their work. In R. J. Holden & R. S. Valdez (Eds.), *The Patient Factor: Theories and Methods for Patient Ergonomics* (pp. 61–81). CRC Press.

Carayon, P., Alvarado, C., & Hundt, A. S. (2003). *Reducing Workload and Increasing Patient Safety Through Work and Workspace Design*. Institute of Medicine.

Carayon, P., Alvarado, C., & Hundt, A. S. (2007a). Work design and patient safety. *TIES-Theoretical Issues in Ergonomics Science*, *8*(5), 395–428.

Carayon, P., Alvarado, C. J., & Hundt, A. S. (2007b). Work system design in health care. In P. Carayon (Ed.), *Handbook of Human Factors and Ergonomics in Healthcare and Patient Safety* (pp. 61–78). Lawrence Erlbaum Associates.

Carayon, P., Alyousef, B., Hoonakker, P., Hundt, A. S., Cartmill, R., Tomcavage, J., Hassol, A., Chaundy, K., Larson, S., Younkin, J., & Walker, J. (2012). Challenges to care coordination posed by the use of multiple health IT applications. *Work: A Journal of Prevention, Assessment and Rehabilitation*, *41*(0), 4468–4473.

Carayon, P., Alyousef, B., & Xie, A. (2012). Human factors and ergonomics in health care. In G. Salvendy (Ed.), *Handbook of Human Factors and Ergonomics* (Fourth ed., pp. 1574–1595). John Wiley & Sons.

Carayon, P., Cartmill, R., Hoonakker, P., Hundt, A. S., Karsh, B.-T., Krueger, D., Snellman, M. L., Thuemling, T. N., & Wetterneck, T. B. (2012). Human factors analysis of workflow in health information technology implementation. In P. Carayon (Ed.), *Handbook of Human Factors and Ergonomics in Health Care and Patient Safety* (Second ed., pp. 507–521). Taylor & Francis Group.

Carayon, P., Cartmill, R., Hoonakker, P., Hundt, A. S., Salek, D., Walker, J., & Tomcavage, J. (2016). Collaborative processes of care managers in the detection and recovery of medication errors. In V. Mollo & P. Falzon (Eds.), *Proceedings of the Healthcare Systems Ergonomics and Patient Safety Conference* (pp. 87–92). IEA Press.

Carayon, P., Faye, H., Hundt, A. S., Karsh, B.-T., & Wetterneck, T. (2011). Patient safety and proactive risk assessment. In Y. Yuehwern (Ed.), *Handbook of Healthcare Delivery Systems* (pp. 11–12). Taylor & Francis Group.

Carayon, P., Grenier, A.-S., & Alvarado, C. (2007c). Macroergonomic and implementation issues of guidelines for accessible medical devices. In J. M. Winters & M. F. Story (Eds.), *Medical Instrumentation - Accessibility and Usability Considerations* (pp. 255–265). CRC Press.

Carayon, P., & Gurses, A. (2005). Nursing workload and patient safety in intensive care units: A human factors engineering evaluation of the literature. *Intensive and Critical Care Nursing*, *21*, 284–301.

Carayon, P., Gurses, A. P., Hundt, A. S., Ayoub, P., & Alvarado, C. J. (2005). Performance obstacles and facilitators of healthcare providers. In C. Korunka & P. Hoffmann (Eds.), *Change and Quality in Human Service Work* (Vol. 4, pp. 257–276). Hampp Publishers.

Carayon, P., Hancock, P., Leveson, N., Noy, I., Sznelwar, L., & van Hootegem, G. (2015). Advancing a sociotechnical systems approach to workplace safety - developing the conceptual framework. *Ergonomics*, *58*(4), 548–564.

Carayon, P., & Hoonakker, P. (2019). Human factors and usability for health information technology: Old and new challenges. *Yearbook of Medical Informatics*, *28*(1), 71–77.

Carayon, P., Hoonakker, P., & Hose, B.-Z. (2025). Human-centered design for patient and clinician safety. In H. M. Kahlid (Ed.), *Helander's Compendium of Human Factors and Ergonomics - Contemporary Issues in Design*. CRC Press.

Carayon, P., Hoonakker, P., Hundt, A. S., Salwei, M., Wiegmann, D., Brown, R. L., Kleinschmidt, P., Novak, C., Pulia, M., Wang, Y., Wirkus, E., & Patterson, B. (2020). Application of human factors to improve usability of clinical decision support for diagnostic decision-making: A scenario-based simulation study. *BMJ Quality & Safety*, *29*, 329–340.

Carayon, P., Hose, B. Z., Wooldridge, A., Brazelton, T. B., Dean, S. M., Eithun, B. L., Kelly, M. M., Kohler, J. E., Ross, J., Rusy, D. A., & Hoonakker, P. L. T. (2022). Human-centered design of team health IT for pediatric trauma care transitions. *International Journal of Medical Informatics*, *162*, 104727.

Carayon, P., Hundt, A. S., Alvarado, C. J., Springman, S., & Ayoub, P. (2006). Patient safety in outpatient surgery: The viewpoint of the healthcare providers. *Ergonomics*, *49*(5–6), 470–485.

Carayon, P., Hundt, A. S., & Hoonakker, P. (2019). Technology barriers and strategies in coordinating care for chronically ill patients. *Applied Ergonomics*, *78*, 240–247.

Carayon, P., Hundt, A. S., Karsh, B.-T., Gurses, A. P., Alvarado, C. J., Smith, M., & Brennan, P. F. (2006). Work system design for patient safety: The SEIPS model. *Quality & Safety in Health Care*, *15*(Supplement I), i50–i58.

Carayon, P., Ju, F., Cartmill, R., Hoonakker, P., Wetterneck, T. B., & Li, J. (2015). Medication error propagation in intensive care units. In The Human Factors and Ergonomics Society (Ed.), *Proceedings of the Human Factors and Ergonomics Society Annual Meeting* (pp. 518–521). The Human Factors and Ergonomics Society.

Carayon, P., Kianfar, S., Li, Y., Xie, A., Alyousef, A., & Wooldridge, A. (2015). A systematic review of mixed methods research on human factors and ergonomics in health care. *Applied Ergonomics*, *51*, 291–321.

Carayon, P., Li, Y., Kelly, M. M., DuBenske, L. L., Xie, A., McCabe, B., Orne, J., & Cox, E. D. (2014). Stimulated recall methodology for assessing work system barriers and facilitators in family-centered rounds in a pediatric hospital. *Applied Ergonomics*, *45*(6), 1540–1546.

Carayon, P., & Perry, S. (2021). Human factors and ergonomics systems approach to the COVID-19 healthcare crisis. *International Journal for Quality in Health Care*, *33*(Supplement_1), 1–3.

Carayon, P., & Salwei, M. E. (2021). Moving toward a sociotechnical systems approach to continuous health information technology design: The path forward for improving electronic health record usability and reducing clinician burnout. *Journal of the American Medical Informatics Association*, *28*(5), 1026–1028.

Carayon, P., Schultz, K., & Hundt, A. S. (2004). Righting wrong site surgery. *Joint Commission Journal on Quality and Safety*, *30*(7), 405–410.

Carayon, P., Smith, M., & Haims, M. (1999). Work organization, job stress, and work-related musculoskeletal disorders. *Human Factors*, *41*(4), 644–663.

Carayon, P., Thuemling, T., Parmasad, V., Bao, S., O'Horo, J., Bennett, N. T., & Safdar, N. (2021). Implementation of an antibiotic stewardship intervention to reduce prescription of fluoroquinolones: A human factors analysis in two intensive care units. *Journal of Patient Safety and Risk Management*, *26*(4), 161–171.

Carayon, P., Wetterneck, T. B., Cartmill, R., Blosky, M. A., Brown, R., Hoonakker, P., Kim, R., Kukreja, S., Johnson, M., Paris, B. L., Wood, K. E., & Walker, J. M. (2021). Medication safety in two intensive care units of a community teaching hospital after electronic health record implementation: Sociotechnical and human factors engineering considerations. *Journal of Patient Safety*, *17*(5), e429–e439.

Carayon, P., Wetterneck, T. B., Cartmill, R., Blosky, M. A., Brown, R., Kim, R., Kukreja, S., Johnson, M., Paris, B., Wood, K. E., & Walker, J. (2014). Characterising the complexity of medication safety using a human factors approach: An observational study in two intensive care units. *BMJ Quality & Safety*, *23*(1), 56–75.

Carayon, P., Wetterneck, T. B., Hundt, A. S., Ozkaynak, M., DeSilvey, J., Ludwig, B., Ram, P., & Rough, S. S. (2007d). Evaluation of nurse interaction with bar code medication administration technology in the work environment. *Journal of Patient Safety*, *3*(1), 34–42.

Carayon, P., Wetterneck, T. B., Hundt, A. S., Rough, S., & Schroeder, M. (2008). Continuous technology implementation in health care: The case of advanced IV infusion pump technology. In K. Zink (Ed.), *Corporate Sustainability As a Challenge for Comprehensive Management* (pp. 139–151). Springer.

Carayon, P., Wetterneck, T. B., Rivera-Rodriguez, A. J., Hundt, A. S., Hoonakker, P., Holden, R., & Gurses, A. P. (2014). Human factors systems approach to healthcare quality and patient safety. *Applied Ergonomics*, *45*(1), 14–25.

Carayon, P., Wooldridge, A., Hoonakker, P., Hundt, A. S., & Kelly, M. M. (2020). SEIPS 3.0: Human-centered design of the patient journey for patient safety. *Applied Ergonomics*, *84*, 103033.

Carayon, P., Wooldridge, A., Hose, B.-T., Salwei, M., & Benneyan, J. (2018). Challenges and opportunities for improving patient safety through human factors and systems engineering. *Health Affairs*, *37*(11), 1862–1869.

Carayon, P., Wust, K., Barton, H. J., Shah, M. N., Dail, P. V., King, B. J., Buckley, D. M., Ehlenfeldt, B., Patterson, B. W., Hoonakker, P., & Werner, N. E. (2025). A mixed methods SEIPS-based evaluation of a patient-centered ED discharge process redesign for older adults. *Journal of Patient Safety*, *21*(4), 203–212.

Carayon, P., Wust, K., Hose, B.-Z., & Salwei, M. (2021). Human factors and ergonomics in health care. In G. Salvendy & W. Karwowski (Eds.), *Handbook of Human Factors and Ergonomics* (Fifth ed., pp. 1417–1437). John Wiley & Sons.

Carayon, P., & Xie, A. (2011). Decision making in healthcare system design: When human factors engineering meets health care. In R. W. Proctor, S. Y. Nof, & Y. Yih (Eds.), *Cultural Factors in Decision Making and Action* (pp. 219–238). Taylor & Francis Group.

Carayon, P., Xie, A., & Kianfar, S. (2013). Human factors and ergonomics. In P. G. Shekelle, R. Wachter, P. Pronovost, K. Schoelles, K. McDonald, S. Dy, K. Shojania, J. Reston, Z. Berger, B. Johnsen, J. Larkin, S. Lucas, K. Martinez, A. Motala, S. Newberry, M. Noble, E. Pfoh, S. Ranji, S. Rennke, E. Schmidt, R. Shanman, N. Sullivan, F. Sun, K. Tipton, J. Treadwell, A. Tsou, M. Vaiana, S. Weaver, R. Wilson, & B. Winters (Eds.), *Making Health Care Safer II: An Updated Critical Analysis of the Evidence for Patient Safety Practices. Comparative Effectiveness Review No. 211* (Vol. AHRQ Publication No.13-E001-EF, pp. 325–350). Agency for Healthcare Research and Quality.

Carayon, P., Xie, A., & Kianfar, S. (2014). Human factors and ergonomics as a patient safety practice. *BMJ Quality & Safety*, *23*, 196–205.

Carey, G., Malbon, E., Carey, N., Joyce, A., Crammond, B., & Carey, A. (2015). Systems science and systems thinking for public health: A systematic review of the field. *BMJ Open Access*, *5*(12), 1–9.

Carman, E.-M., Fray, M., & Waterson, P. (2021). Facilitators and barriers of care transitions: Comparing the perspectives of hospital and community healthcare staff. *Applied Ergonomics*, *93*, 103339.

Castro, E. M., Van Regenmortel, T., Vanhaecht, K., Sermeus, W., & Van Hecke, A. (2016). Patient empowerment, patient participation and patient-centeredness in hospital care: A concept analysis based on a literature review. *Patient Education and Counseling*, *99*(12), 1923–1939.

Catchpole, K., Bowie, P., Fouquet, S., Rivera, J., & Hignett, S. (2021). Frontiers in human factors: Embedding specialists in multi-disciplinary efforts to improve healthcare. *International Journal for Quality in Health Care*, *33*(Supplement_1), 13–18.

Catchpole, K., & Wiegmann, D. (2012). Understanding safety and performance in the cardiac operating room: From 'sharp end' to 'blunt end'. *BMJ Quality & Safety*, *21*(10), 807–809.

CDC. (2019). *Core Elements of Hospital Antibiotic Stewardship Programs*. US Department of Health and Human Services, CDC.

Centre for Transformative Work Design. (2025). *SMART work design*. Retrieved April 2, 2025 from https://www.transformativeworkdesign.com/smart-work

Chapanis, A., & Safrin, M. A. (1960). Of misses and medicines. *Journal of Chronic Diseases*, *12*(4), 403–408.

Chari, R., Chang, C.-C., Sauter, S. L., Petrun Sayers, E. L., Cerully, J. L., Schulte, P., Schill, A. L., & Uscher-Pines, L. (2018). Expanding the paradigm of occupational safety and health: A new framework for worker well-being. *Journal of Occupational and Environmental Medicine*, *60*(7), 589–593.

Chari, R., Sauter, S. L., Petrun Sayers, E. L., Huang, W., Fisher, G. G., & Chang, C. C. (2022). Development of the national institute for occupational safety and health worker well-being questionnaire. *Journal of Occupational and Environmental Medicine*, *64*(8), 707–717.

Checkland, P. (1999). *Systems Thinking, Systems Practice*. John Wiley & Sons.

Cherns, A. (1976). The principles of sociotechnical design. *Human Relations*, *29*(8), 783–792.

Cherns, A. (1987). Principles of sociotechnical design revisited. *Human Relations*, *40*(3), 153–162.

Choi, Y. S., Yi, J. S., Law, C. M., & Jacko, J. A. (2006). *Are "Universal Design Resources" designed for designers?* ASSETS'06, Portland, Oregon, USA.

Chui, M. A., Mott, D. A., & Maxwell, L. (2012). A qualitative assessment of a community pharmacy cognitive pharmaceutical services program, using a work system approach. *Research in Social & Administrative Pharmacy*, *8*(3), 206–216.

Clark, D., Lawton, R., Baxter, R., Sheard, L., & O'Hara, J. K. (2025). Do healthcare professionals work around safety standards, and should we be worried? A scoping review. *BMJ Quality & Safety*, *34*(5), 317–329.

Clarkson, J., Bogle, D., Dean, J., Tooley, M., Trewby, J., Vaughan, L., Adams, E., Dudgeon, P., Platt, N., & Shelton, P. (2017). *Engineering Better Care: A Systems Approach to Health and Care Design and Continuous Improvement*. Royal Academy of Engineering.

Clay-Williams, R., Hounsgaard, J., & Hollnagel, E. (2015). Where the rubber meets the road: Using FRAM to align work-as-imagined with work-as-done when implementing clinical guidelines. *Implementation Science*, *10*(1), 1–8.

Clegg, C. W. (2000). Sociotechnical principles for system design. *Applied Ergonomics*, *31*, 463–477.

Cohen, T. N., Francis, S. E., Wiegmann, D. A., Shappell, S. A., & Gewertz, B. L. (2018). Using HFACS-Healthcare to identify systemic vulnerabilities during surgery. *American Journal of Medical Quality*, *33*(6), 614–622.

Cohen, W. J., & Smith, J. H. (1994). Community ergonomics: Past approaches and future prospects toward America's urban crisis. In The Human Factors and Ergonomics Society (Ed.), *Proceedings of the Human Factors and Ergonomics Society 38th Annual Meeting* (pp. 734–738). The Human Factors and Ergonomics Society.

Coiera, E. (2011). Why system inertia makes health reform so difficult. *BMJ*, *342*, d3693.

Cook, R. I., Woods, D. D., & Miller, C. (1998). *A Tale of Two Stories: Contrasting Views of Patient Safety*. National Patient Safety Foundation.

Cooper, M. D. (2020). The emperor has no clothes: A critique of safety-II. *Safety Science*, *152*, 105047.

Coppola, S. M., Neyens, D. M., Catchpole, K., Gurses, A. P., Osei, P., Biro, J. M., Alfred, M., Rucks, M., Tobin, C. D., Jaruzal, C., Abernathy, J. H., & Operating Room System based Medication Administration error Reduction Team Research Group. (2022). Systems engineering-based framework of process risks in perioperative medication delivery. *British Journal of Anaesthesia*, *129*(6), e168–e170.

Cox, E. D., Jacobson, G. C., Rajamanickam, V. P., Carayon, P., Kelly, M. M., Wetterneck, T. B., Rathouz, P. J., & Brown, R. L. (2017). A family-centered rounds checklist, family engagement, and patient safety: A randomized trial. *Pediatrics*, *139*(5), e20161688.

Cullen, D. J., Sweitzer, B. J., Bates, D. W., Burdick, E., Edmondson, A., & Leape, L. L. (1997). Preventable adverse drug events in hospitalized patients: A comparative study of intensive care and general care units. *Critical Care Medicine*, *25*(8), 1289–1297.

Cummings, T. G. (1978). Self-regulating work groups: A socio-technical synthesis. *Academy of Management Review*, *3*(3), 625–634.
Daniellou, F. (2005). The French-speaking ergonomists' approach to work activity: Cross-influences of field intervention and conceptual models. *Theoretical Issues in Ergonomics Science*, *6*(5), 409–427.
Davis, K. G., & Kotowski, S. E. (2015). Prevalence of musculoskeletal disorders for nurses in hospitals, long-term care facilities, and home health care: A comprehensive review. *Human Factors*, *57*(5), 754–792.
Davis, L. E. (1966). The design of jobs. *Industrial Relations Society*, *6*(1), 21–45.
Davis, L. E. (1977). Evolving alternative organization designs: Their sociotechnical bases. *Human Relations*, *30*(3), 261–273.
Davis, L. E., & Wacker, G. J. (1987). Job design. In G. Salvendy (Ed.), *Handbook of Human Factors* (pp. 431–452). John Wiley & Sons.
Davis, M. C., Challenger, R., Jayewardene, D. N., & Clegg, C. W. (2014). Advancing socio-technical systems thinking: A call for bravery. *Applied Ergonomics*, *45*(2), 171–180.
De Bienassis, K., Slawomirski, L., & Klazinga, N. S. (2021). *The Economics of Patient Safety Part IV: Safety in the Workplace*. OECD
De Ridder, E., Dekkers, T., Porsius, J. T., Kraan, G., & Melles, M. (2018). The perioperative patient experience of hand and wrist surgical patients: An exploratory study using patient journey mapping. *Patient Experience Journal*, *5*, 97–107).
De Zwart, F. (2015). Unintended but not unanticipated consequences. *Theory and Society*, *44*(3), 283–297.
Dekker, S. (2011). *Patient Safety - A Human Factors Approach*. CRC Press.
Dekker, S. (2014). *The Field Guide to Understanding 'Human Error'* (Third ed.). CRC Press.
Dekker, S. W., & Leveson, N. G. (2014). The bad apple theory won't work: Response to 'Challenging the systems approach: why adverse event rates are not improving' by Dr Levitt. *BMJ Quality & Safety*, *23*(12), 1050–1051.
Dekker, S. W., & Leveson, N. G. (2015). The systems approach to medicine: Controversy and misconceptions. *BMJ Quality & Safety*, *24*(1), 7–9.
Demerouti, E. (2014). Design your own job through job crafting. *European Psychologist*, *19*(4), 237–247.
Derlet, R. W. (2002). Overcrowding in emergency departments: Increased demand and decreased capacity. *Annals of Emergency Medicine*, *39*(4), 430–432.
Design Council. (2025). *The Double Diamond*. Retrieved February 6, 2025 from https://www.designcouncil.org.uk/our-resources/the-double-diamond/
Détienne, F. (2006). Collaborative design: Managing task interdependencies and multiple perspectives. *Interacting with Computers*, *18*(1), 1–20.
Dixon-Woods, M. (2010). Why is patient safety so hard? A selective review of ethnographic studies. *Journal of Health Services Research and Policy*, *15*, 11–16.
Dixon-Woods, M. (2014). *The Problem of Context in Quality Improvement*. The Health Foundation.
Dixon-Woods, M., Bosk, C. L., Aveling, E. L., Goeschel, C. A., & Pronovost, P. J. (2011). Explaining Michigan: Developing an ex post theory of a quality improvement program. *The Milbank Quarterly*, *89*(2), 167–205.
Dixon-Woods, M., McNicol, S., & Martin, G. (2012). Ten challenges in improving quality in healthcare: Lessons from the Health Foundation's programme evaluations and relevant literature. *BMJ Quality & Safety*, *21*(10), 876–884.
Dixon-Woods, M., & Pronovost, P. J. (2016). Patient safety and the problem of many hands. *BMJ Quality & Safety*, *25*(7), 485–488.
Donabedian, A. (1966). Evaluating the quality of medical care. *Millbank Memorial Fund Quarterly*, *40*(3), 166–206.

Donabedian, A. (1978). The quality of medical care. *Science*, *200*, 856–864.

Dressner, M. A. (2017). Hospital workers: An assessment of occupational injuries and illnesses. *Monthly Labor Review*, *140*, 1.

Dressner, M. A., & Kissinger, S. P. (2018). Occupational injuries and illnesses among registered nurses. *Monthly Labor Review*, *141*, 1.

Dul, J., Bruder, R., Buckle, P., Carayon, P., Falzon, P., Marras, W. S., Wilson, J. R., & van der Doelen, B. (2012). A strategy for human factors/ergonomics: Developing the discipline and profession. *Ergonomics*, *55*(4), 377–395.

Duncan, R., & Weiss, A. (1979). Organizational learning: Implications for organizational learning. *Research in Organizational Behavior*, *1*, 75–123.

Durso, F. T., Ferguson, A. N., Kazi, S., Cunningham, C., & Ryan, C. (2014). Strategic threat management: An exploration of nursing strategies in the pediatric intensive care unit. *Applied Ergonomics*, *47*, 345–354.

Eason, K. (1995). User-centred design: For users or by users? *Ergonomics*, *38*(8), 1667–1673.

Eason, K. (2014). Afterword: The past, present and future of sociotechnical systems theory. *Applied Ergonomics*, *45*(2, Part A), 213–220.

Edmondson, A. (1999). Psychological safety and learning behavior in work teams. *Administrative Science Quarterly*, *44*(2), 350–383.

Eklund, J. H., Holmstrom, I. K., Kumlin, T., Kaminsky, E., Skoglund, K., Hoglander, J., Sundler, A. J., Conden, E., & Meranius, M. S. (2019). "Same same or different?" A review of reviews of person-centered and patient-centered care. *Patient Education and Counseling*, *102*(1), 3–11.

Ellapen, T. J., & Narsigan, S. (2015). Work related musculoskeletal disorders among nurses: Systematic review. *Journal of Ergonomics*, *4*, S4-003.

Elseviers, M. M., Arias-Guillen, M., Gorke, A., & Arens, H. J. (2014). Sharps injuries amongst healthcare workers: Review of incidence, transmissions and costs. *Journal of Renal Care*, *40*(3), 150–156.

Emery, F., & Trist, E. (1965). The causal texture of organizational environments. *Human Relations*, *18*, 21–31.

Emery, M., & Purser, R. E. (1996). *The Search Conference - A Powerful Method for Planning Organizational Change and Community Action*. Jossey-Bass Publishers.

Endsley, M. R. (2023). Ironies of artificial intelligence. *Ergonomics*, *66*(11), 1656–1668.

English, M., Ogola, M., Aluvaala, J., Gicheha, E., Irimu, G., McKnight, J., & Vincent, C. A. (2021). First do no harm: Practitioners' ability to 'diagnose' system weaknesses and improve safety is a critical initial step in improving care quality. *Archives of Disease in Childhood*, *106*(4), 326–332.

Etchgaray, J. M., & Thomas, E. J. (2014). Engaging employees: The importance of high-performance work systems for patient safety. *Journal of Patient Safety*, *11*, 211–227.

Evanoff, V. A., Bohr, P. C., & Wolf, L. (1999). Effects of a participatory ergonomics team among hospital orderlies. *American Journal of Industrial Medicine*, *35*, 358–365.

Eyal, Z., Daniel, H., Orly, T., Calvin, F., Peter, S., Cyrus, K. Y., Carol, K., Charles, R. D., & David, W. B. (2013). Health care-associated infections: A meta-analysis of costs and financial impact on the US health care system. *JAMA Internal Medicine*, *173*(22), 2039–2046.

Falzon, P. (2007). Enabling safety: Issues in design and continuous design. *Cognition, Technology & Work*, *10*(1), 7–14.

Falzon, P. (Ed.). (2015). *Constructive Ergonomics*. CRC Press.

Faye, H., & Falzon, P. (2009). Strategies of performance self-monitoring in automotive production. *Applied Ergonomics*, *40*(5), 915–921.

Faye, H., Rivera-Rodriguez, A. J., Karsh, B. T., Hundt, A. S., Baker, C., & Carayon, P. (2010). Involving intensive care unit nurses in a proactive risk assessment of the medication management process. *The Joint Commission Journal on Quality and Patient Safety*, *36*(8), 376–384.

Fell, S. E., Bowie, P., Spalding, K., & Medves, J. (2024). Preliminary adaptation of the systems thinking for everyday work cue card set in a US healthcare system: A pragmatic and participatory co-design approach. *BMJ Open Quality*, *13*(2), e002655.

Flach, J. M. (2012). Complexity: Learning to muddle through. *Cognition, Technology & Work*, *14*, 187–197.

Flanagan, J. C. (1954). The critical incident technique. *Psychological Bulletin*, *51*(4), 327–358.

Flynn, E. A. (2012). Human factors and ergonomics in medication safety. In P. Carayon (Ed.), *Handbook of Human Factors and Ergonomics in Health Care and Patient Safety* (Second ed., pp. 785–792). Taylor & Francis Group.

Forrester, J. D., Maggio, P. M., & Tennakoon, L. (2022). Cost of health care-associated infections in the United States. *Journal of Patient Safety*, *18*, 477–479.

Forrester, J. W. (2007a). System dynamics: A personal view of the first fifty years. *System Dynamics Review*, *23*(2–3), 345–358.

Forrester, J. W. (2007b). System dynamics: The next fifty years. *System Dynamics Review*, *23*(2–3), 359–370.

Fouquet, S. D., Fitzmaurice, L., Chan, Y. R., & Palmer, E. M. (2020). Doctors documenting: An ethnographic and informatics approach to understanding attending physician documentation in the pediatric emergency department. *Journal of the American Medical Informatics Association*, *28*(2), 239–248.

Fox, W. M. (1995). Sociotechnical system principles and guidelines: Past and present. *Journal of Applied Behavioral Science*, *31*(1), 91–105.

Frankel, A., Graydon-Baker, E., Neppl, C., Simmonds, T., Gustafson, M., & Gandhi, T. K. (2003). Patient safety leadership walkrounds™. *The Joint Commission Journal on Quality and Safety*, *29*(1), 16–26.

Fraser, S. W., & Greenhalgh, T. (2001). Coping with complexity: Educating for capability. *BMJ*, *323*(7316), 799–803.

Fujishiro, K., Weaver, J. L., Heaney, C. A., Hamrick, C. A., & Marras, W. S. (2005). The effect of ergonomic interventions in healthcare facilities on musculoskeletal disorders. *American Journal of Industrial Medicine*, *48*(5), 338–347.

Gates, P. J., Hardie, R. A., Raban, M. Z., Li, L., & Westbrook, J. I. (2021). How effective are electronic medication systems in reducing medication error rates and associated harm among hospital inpatients? A systematic review and meta-analysis. *Journal of the American Medical Informatics Association*, *28*(1), 167–176.

Gerain, P., & Zech, E. (2019). Informal caregiver burnout? Development of a theoretical framework to understand the impact of caregiving. *Frontiers in Psychology*, *10*, 1–13.

Gilson, A. M., Xiong, K. Z., Stone, J. A., Jacobson, N., & Chui, M. A. (2021). A pharmacy-based intervention to improve safe over-the-counter medication use in older adults. *Research in Social and Administrative Pharmacy*, *17*(3), 578–587.

Göttgens, I., & Oertelt-Prigione, S. (2021). The application of human-centered design approaches in health research and innovation: A narrative review of current practices. *JMIR Mhealth and Uhealth*, *9*(12), e28102.

Goyal, A., Glanzman, H., Quinn, M., Tur, K., Singh, S., Winter, S., Snyder, A., & Chopra, V. (2020). Do bedside whiteboards enhance communication in hospitals? An exploratory multimethod study of patient and nurse perspectives. *BMJ Quality & Safety*, *29*(10), 1–2.

Graber, M. L. (2013). The incidence of diagnostic error in medicine. *BMJ Quality & Safety*, *22*(Suppl 2), ii21–ii27.

Graber, M. L., Castro, G. M., Danforth, M., Tilly, J. L., Croskerry, P., El-Kareh, R., Hemmalgarn, C., Ryan, R., Tozier, M. P., Trowbridge, B., Wright, J., & Zwaan, L. (2024). Root cause analysis of cases involving diagnosis. *Diagnosis*, *11*(4), 353–368.

Graber, M. L., Franklin, N., & Gordon, R. (2005). Diagnostic error in internal medicine. *Archives of Internal Medicine*, *165*(13), 1493–1499.

Grant, R., Benamouzig, D., Catton, H., Cheng, V. C., Dhingra, N., Laxminarayan, R., Legido-Quigley, H., Martinez, S. L., Marikar, K., Mendelson, M., Parneix, P., Perencevich, E., Singh, H., Tschudin-Sutter, S., Ushiro, S., Wesangula, E., Gardiol, C., Dziekan, G., Harbarth, S., & Pittet, D. (2023). COVID-19 pandemic: A catalyst for accelerating global action on patient safety. *Lancet Infectious Diseases*, *23*(10), 1108–1110.

Greenhalgh, J., & Manzano, A. (2022). Understanding 'context' in realist evaluation and synthesis. *International Journal of Social Research Methodology*, *25*(5), 583–595.

Greenhalgh, T. (2014). *How to Read a Paper - The Basics of Evidence-Based Medicine* (Fifth ed.). Wiley Blackwell.

Greenhalgh, T. (2018). *How to Implement Evidence-Based Healthcare*. John Wiley & Sons.

Greenhalgh, T., & Papoutsi, C. (2018). Studying complexity in health services research: Desperately seeking an overdue paradigm shift. *BMC Medicine*, *16*(1), 95.

Greenhalgh, T., Robert, G., MacFarlane, F., Bate, P., & Kyriakidou, O. (2004). Diffusion of innovations in service organizations: Systematic review and recommendations. *The Milbank Quarterly*, *82*(4), 581–629.

Grimes, T. C. (2021). Is it time for greater patient involvement to enhance transitional medication safety? *BMJ Quality & Safety*, *31*(4), 247–250.

Grindell, C., Coates, E., Croot, L., & O'Cathain, A. (2022). The use of co-production, co-design and co-creation to mobilise knowledge in the management of health conditions: A systematic review. *BMC Health Services Research*, *22*(1), 877.

Grover, S., Fitzpatrick, A., Azim, F. T., Ariza-Vega, P., Bellwood, P., Burns, J., Burton, E., Fleig, L., Clemson, L., Hoppmann, C. A., Madden, K. M., Price, M., Langford, D., & Ashe, M. C. (2022). Defining and implementing patient-centered care: An umbrella review. *Patient Education and Counseling*, *105*(7), 1679–1688.

Guerin, F., Laville, A., Daniellou, F., Duraffourg, J., & Kerguelen, A. (2006). *Understanding and Transforming Work - The Practice of Ergonomics*. ANACT.

Guest, J. F., Keating, T., Gould, D., & Wigglesworth, N. (2020). Modelling the annual NHS costs and outcomes attributable to healthcare-associated infections in England. *BMJ Open*, *10*(1), e033367.

Gupta, A., Harrod, M., Quinn, M., Manojlovich, M., Fowler, K. E., Singh, H., Saint, S., & Chopra, V. (2018). Mind the overlap: How system problems contribute to cognitive failure and diagnostic errors. *Diagnosis*, *5*(3), 151–156.

Gurses, A., & Carayon, P. (2009). Exploring performance obstacles of intensive care nurses. *Applied Ergonomics*, *40*(3), 509–518.

Gurses, A. P., & Carayon, P. (2007). Performance obstacles of intensive care nurses. *Nursing Research*, *56*(3), 185–194.

Gurses, A. P., Carayon, P., & Wall, M. (2009). Impact of performance obstacles on intensive care nurses workload, perceived quality and safety of care, and quality of working life. *Health Services Research*, *44*(2), 422–443.

Gurses, A. P., Dietz, A. S., Nowakowski, E., Andonian, J., Schiffhauer, M., Billman, C., Abashian, A. M., Trexler, P., Osei, P., Benishek, L. E., Xie, A., Pronovost, P., Rosen, M. A., & Maragakis, L. L. (2019). Human factors–based risk analysis to improve the safety of doffing enhanced personal protective equipment. *Infection Control & Hospital Epidemiology*, *40*(2), 178–186.

Gurses, A. P., Kim, G., Martinez, E. A., Marsteller, J., Bauer, L., Lubomski, L. H., Pronovost, P. J., & Thompson, D. (2012). Identifying and categorising patient safety hazards in cardiovascular operating rooms using an interdisciplinary approach: A multisite study. *BMJ Quality & Safety*, *21*(10), 810–818.

Gurses, A. P., Ozok, A. A., & Pronovost, P. J. (2012). Time to accelerate integration of human factors and ergonomics in patient safety. *BMJ Quality & Safety*, *21*(4), 347–351.

Gurses, A. P., Seidl, K. L., Vaidya, V., Bochicchio, G., Harris, A. D., Hebden, J., & Xiao, Y. (2008). Systems ambiguity and guideline compliance: A qualitative study of how intensive care units follow evidence-based guidelines to reduce healthcare-associated infections. *Quality and Safety in Health Care, 17*(5), 351–359.

Hackman, J. R., & Oldham, G. R. (1976). Motivation through the design of work: Test of a theory. *Organizational Behavior and Human Performance, 16*(2), 250–279.

Hackman, J. R., Oldham, G. R., Janson, R., & Purdy, K. (1975). A new strategy for job enrichment. *California Management Review, 17*(4), 57–71.

Haines, H., Wilson, J. R., Vink, P., & Koningsveld, E. (2002). Validating a framework for participatory ergonomics (the PEF). *Ergonomics, 45*(4), 309–327.

Hall, L. H., Johnson, J., Watt, I., Tsipa, A., & O'Connor, D. B. (2016). Healthcare staff wellbeing, burnout, and patient safety: A systematic review. *PloS One, 11*(7), e0159015.

Hall-Andersen, L. B., & Broberg, O. (2014). Integrating ergonomics into engineering design: The role of objects. *Applied Ergonomics, 45*(3), 647–654.

Hallock, M. L., Alper, S. J., & Karsh, B. (2006). A macroergonomic work system analysis of the diagnostic testing process in an outpatient health care facility for process improvement and patient safety. *Ergonomics, 49*(5–6), 544–566.

Hamilton-Escoto, K., Hallock, M., Wagner, J., & Karsh, B. (2004). Using variance analysis to detect hazards in a bar-code assisted medication preparation process. *Joint Commission Journal on Quality and Safety, 30*(11), 622–628.

Hansen, A. F., Hasle, P., Caroly, S., Reinhold, K., Jarvis, M., Herrig, A. O., Heiberg, B. D., Sogaard, K., Punnett, L., & Jensen Stochkendahl, M. (2024). Participatory ergonomics: What works for whom and why? A realist review *Ergonomics, 67*(1), 13–33.

Hay, G. J., Klonek, F. E., & Parker, S. K. (2020). Diagnosing rare diseases: A sociotechnical approach to the design of complex work systems. *Applied Ergonomics, 86*, 103095.

Hay, G. J., Klonek, F. E., Thomas, C. S., Bauskis, A., Baynam, G., & Parker, S. K. (2020). SMART work design: Accelerating the diagnosis of rare diseases in the Western Australian undiagnosed diseases program. *Frontiers in Pediatrics, 8*(582).

Hendrick, H. W. (1991). Ergonomics in organizational design and management. *Ergonomics, 34*, 743–756.

Hendrick, H. W., & Kleiner, B. M. (2001). *Macroergonomics: An Introduction to Work System Design*. The Human Factors and Ergonomics Society.

Hesgrove, B., Zebrak, K., Yount, N., Sorra, J., & Ginsberg, C. (2024). Associations between patient safety culture and workplace safety culture in hospital settings. *BMC Health Services Research, 24*(1), 568.

Hignett, S., Carayon, P., Buckle, P., & Catchpole, K. (2013). State of science: Human factors and ergonomics in healthcare. *Ergonomics, 56*(10), 1491–1503.

Hignett, S., & Lu, J. (2010). Space to care and treat safely in acute hospitals: Recommendations from 1866 to 2008. *Applied Ergonomics, 41*(5), 666–673.

Hignett, S., & Wolf, L. (2016). Reducing inpatient falls: Human Factors & Ergonomics offers a novel solution by designing safety from the patients' perspective. *International Journal of Nursing Studies, 59*, A1–A3.

Holden, R. J., & Carayon, P. (2021). SEIPS 101 and seven simple SEIPS tools. *BMJ Quality & Safety, 30*(11), 901–910.

Holden, R. J., Carayon, P., Gurses, A. P., Hoonakker, P., Hundt, A. S., Ozok, A. A., & Rivera-Rodriguez, A. J. (2013). SEIPS 2.0: A human factors framework for studying and improving the work of healthcare professionals and patients. *Ergonomics, 56*(11), 1669–1686.

Holden, R. J., Cornet, V. P., & Valdez, R. S. (2020). Patient ergonomics: 10-year mapping review of patient-centered human factors. *Applied Ergonomics, 82*, 102972.

Holden, R. J., Schubert, C. C., Eiland, E. C., Storrow, A. B., Miller, K. F., & Collins, S. P. (2015). Self-care barriers reported by emergency department patients with acute heart failure: A sociotechnical systems-based approach. *Annals of Emergency Medicine*, *66*(1), 1–12.

Holden, R. J., Schubert, C. C., & Mickelson, R. S. (2015). The patient work system: An analysis of self-care performance barriers among elderly heart failure patients and their informal caregivers. *Applied Ergonomics*, *47*(0), 133–150.

Holden, R. J., & Valdez, R. S. (Eds.). (2021). *The Patient Factor: Theories and Methods for Patient Ergonomics*. CRC Press.

Holden, R. J., Valdez, R. S., Schubert, C. C., Thompson, M. J., & Hundt, A. S. (2017). Macroergonomic factors in the patient work system: Examining the context of patients with chronic illness. *Ergonomics*, *60*(1), 26–43.

Hollnagel, E. (2011). Epilogue: RAG - The resilience analysis grid. In E. Hollnagel, J. Paries, D. D. Woods, & J. Wreathall (Eds.), *Resilience Engineering in Practice - A Guidebook* (pp. 275–296). Ashgate Publishing

Hollnagel, E. (2012a). Coping with complexity: Past, present and future. *Cognition, Technology & Work*, *14*, 199–205.

Hollnagel, E. (2012b). *FRAM: The Functional Resonance Analysis Method: Modelling Complex Socio-technical Systems*. Ashgate Publishing.

Hollnagel, E. (2014). *Safety-I and Safety-II - The Past and Future of Safety Management*. Ashgate Publishing.

Hollnagel, E., Braithwaite, J., & Wears, R. L. (2015). *Resilient Health Care*. Ashgate Publishing

Hollnagel, E., & Woods, D. D. (2006). Epilogue: Resilience engineering concepts. In E. Hollnagel, D. D. Woods, & N. Leveson (Eds.), *Resilience Engineering - Concepts and Precepts* (pp. 346–358). Ashgate Publishing

Hollnagel, E., Woods, D. D., & Leveson, N. (Eds.). (2006). *Resilience Engineering: Concepts and Precepts*. Ashgate Publishing.

Hoonakker, P. L., Carayon, P., McGuire, K., Khunlertkit, A., Wiegmann, D. A., Alyousef, B., Xie, A., & Wood, K. E. (2013). Motivation and job satisfaction of Tele-ICU nurses. *Journal of Critical Care*, *28*(3), 315.e313–315.e321.

Hoonakker, P. L. T., & Carayon, P. (2018). Work system barriers and strategies reported by tele-intensive care unit nurses. *Critical Care Nursing Clinics of North America*, *30*(2), 259–271.

Hoonakker, P. L. T., Carayon, P., & Cartmill, R. S. (2017). The impact of secure messaging on workflow in primary care: Results of a multiple-case, multiple-method study. *International Journal of Medical Informatics*, *100*, 63–76.

Hoonakker, P. L. T., Carayon, P., Hose, B.-Z., Ramsey, J. C., Eithun, B. L., Kim, M. K., Koffarnus, K. S., Kohler, J. E., Nieman, J. A., Reisman, M. M., Ross, J. C., Rusy, D. A., & Soetenga, D. J. (2019). Proactive risk assessment of team health IT for pediatric trauma care transitions (T^3). In T. P. Cotrim, F. Serranheira, P. Sousa, S. Hignett, S. Albolino, & R. Tartaglia (Eds.), *Health and Social Care Systems of the Future: Demographic Changes, Digital Age and Human Factors* (pp. 213–221). Springer International Publishing.

Hoonakker, P. L. T., Hose, B.-Z., Carayon, P., Eithun, B. L., Rusy, D. A., Ross, J. C., Kohler, J. E., Dean, S. M., Brazelton, T. B., & Kelly, M. M. (2022). Scenario-based evaluation of team health information technology to support pediatric trauma care transitions. *Applied Clinical Informatics*, *13*(01), 218–229.

Hoonakker, P. L. T., Wooldridge, A. R., Hose, B. Z., Carayon, P., Eithun, B., Brazelton, T. B., Kohler, J. E., Ross, J. C., Rusy, D. A., Dean, S. M., Kelly, M. M., & Gurses, A. P. (2019). Information flow during pediatric trauma care transitions: Things falling through the cracks. *Internal and Emergency Medicine*, *14*(5), 797–805.

Horgan, S., Hegarty, J., Drennan, J., Keane, D., & Saab, M. M. (2024). The effect of interventions on the incidence of surgical site infections in acute care settings: A systematic review. *Journal of Tissue Viability, 33*(1), 75–88.

Hose, B.-Z., Carayon, P., Hoonakker, P. L. T., Brazelton, T. B., Dean, S. M., Eithun, B. L., Kelly, M. M., Kohler, J. E., Ross, J. C., & Rusy, D. A. (2023). Work system barriers and facilitators of a team health information technology. *Applied Ergonomics, 113*, 104105.

Hose, B.-Z., Carayon, P., Hoonakker, P. L. T., Ross, J. C., Eithun, B. L., Rusy, D. A., Kohler, J. E., Brazelton, T. B., Dean, S. M., & Kelly, M. M. (2023). Managing multiple perspectives in the collaborative design process of a team health information technology. *Applied Ergonomics, 106*, 103846.

Hundt, A. S. (2012). Organizational learning in health care. In P. Carayon (Ed.), *Handbook of Human Factors and Ergonomics in Health Care and Patient Safety* (Second ed., pp. 97–108). Taylor & Francis Group.

Hundt, A. S., Adams, J. A., Schmid, A., Musser, L. M., Walker, J. M., Wetterneck, T. B., Paris, B. L., Douglas, S. V., & Carayon, P. (2013). Conducting an efficient proactive risk assessment prior to CPOE implementation. *International Journal of Medical Informatics, 82*, 25–38.

Hundt, A. S., Carayon, P., Hoonakker, P., Cartmill, R., Den Herder, R., Topper, J., Younkin, J., & Walker, J. (2012). Organizational learning in a large-scale complex health IT project. In The Human Factors and Ergonomics Society (Ed.), *Proceedings of the Human Factors and Ergonomics Society 56th Annual Meeting* (pp. 1288–1292). The Human Factors and Ergonomics Society.

Hundt, A. S., Carayon, P., Wetterneck, T. B., Love, T., Haack, B., Schroeder, M., & Enloe, M. (2005). Evaluating design changes of a smart IV pump. In R. Tartaglia, S. Bagnara, T. Bellandi, & S. Albolino (Eds.), *Healthcare Systems Ergonomics and Patient Safety* (pp. 239–242). Taylor & Francis Group.

Hundt, A. S., Carayon, P., Yang, Y., Stamm, J., Agrawal, V., Kleinschmidt, P., & Hoonakker, P. (2017). Role network analysis of team interactions and individual activities: Application to VTE prophylaxis. In The Human Factors and Ergonomics Society (Ed.), *Proceedings of the Human Factors and Ergonomics Society Annual Meeting* (Vol. 61, pp. 896–900). The Human Factors and Ergonomics Society.

Hutchins, E. (1995). *Cognition in the Wild*. The MIT Press.

Hutchins, E. (2001). Cognition, distributed. In N. J. Smelser & P. B. Baltes (Eds.), *International Encyclopedia of the Social & Behavioral Sciences* (Vol. 3, pp. 2068–2072). Elsevier.

Ibrahim Shire, M., Jun, G. T., Moon, S., & Robinson, S. (2019). A system dynamics approach to workload management of hospital pharmacy staff: Modeling the tradeoff between dispensing backlog and dispensing errors. *IISE Transactions on Occupational Ergonomics and Human Factors, 6*(3–4), 209–224.

Ibrahim Shire, M., Jun, G. T., & Robinson, S. (2018). The application of system dynamics modelling to system safety improvement: Present use and future potential. *Safety Science, 106*, 104–120.

Ibrahim Shire, M., Jun, G. T., & Robinson, S. (2020). Healthcare workers' perspectives on participatory system dynamics modelling and simulation: Designing safe and efficient hospital pharmacy dispensing systems together. *Ergonomics, 63*(8), 1044–1056.

Iflaifel, M., Lim, R. H., Ryan, K., & Crowley, C. (2020). Resilient health care: A systematic review of conceptualisations, study methods and factors that develop resilience. *BMC Health Services Research, 20*, 1–21.

Illingworth, J., Shaw, A., Fernandez Crespo, R., Leis, M., Fontana, G., Howitt, P., & Darzi, A. (2023). *Global State of Patient Safety 2023*. Imperial College London. Retrieved November 25, 2025 from https://www.imperial.ac.uk/Stories/global-state-of-patient-safety/

Institute of Medicine. (2001). *Crossing the Quality Chasm: A New Health System for the 21st Century*. The National Academies Press.

Institute of Medicine. (2004). *Keeping Patients Safe: Transforming the Work Environment of Nurses*. The National Academies Press.

Institute of Medicine. (2012). *Health IT and Patient Safety: Building Safer Systems for Better Care*. The National Academies Press.

International Ergonomics Association (IEA). (2000). *What is ergonomics (HFE)?* Retrieved April 30, 2024 from https://iea.cc/about/what-is-ergonomics/

International Organization for Standardization (ISO). (2010). *ISO 9241-210 Ergonomics of Human-System interaction - Part 210: Human-Centred Design for Interactive Systems*. ISO.

Isherwood, P., & Waterson, P. (2021). To err is system; a comparison of methodologies for the investigation of adverse outcomes in healthcare. *Journal of Patient Safety and Risk Management*, *26*, 64–73.

Jackson, S. L., Read, G. J. M., Hulme, A., & Salmon, P. M. (2025). Systems Human Factors and Ergonomics methods: Applications, outcomes, and future directions. *Human Factors and Ergonomics in Manufacturing & Service Industries*, *35*(3), e70002.

Jamshidi, M. (2009). Introduction to system of systems. In M. Jamshidi (Ed.), *System of Systems Engineering - Innovations for the 21st Century* (pp. 1–20). John Wiley & Sons.

Johns, G. (2006). The essential impact of context on organizational behavior. *Academy of Management Review*, *31*(2), 386–408.

Jolliff, A., Holden, R. J., Valdez, R., Coller, R. J., Patel, H., Zuraw, M., Linden, A., Ganci, A., Elliott, C., & Werner, N. E. (2024). Investigating the best practices for engagement in remote participatory design: Mixed methods analysis of 4 remote studies with family caregivers. *Journal of Medical Internet Research*, *26*, e60353.

Jones, A., Neal, A., Bailey, S., & Cooper, A. (2024). When work harms: How better understanding of avoidable employee harm can improve employee safety, patient safety and healthcare quality. *BMJ Leader*, *8*(1), 59–62.

Jones, C. E. L., Phipps, D. L., & Ashcroft, D. M. (2018). Understanding procedural violations using Safety-I and Safety-II: The case of community pharmacies. *Safety Science*, *105*, 114–120.

Jones, P. H. (2013). *Design for Care - Innovating Healthcare Experience*. Rosenfield Media.

Joseph, A., Henriksen, K., & Malone, E. (2018). The architecture of safety: An emerging priority for improving patient safety. *Health Affairs*, *37*(11), 1884–1891.

Jun, G. T., Hignett, S., & Clarkson, P. J. (2024). *Design Creativity*. Cambridge University Press.

Jun, G. T., Ward, J., Morris, Z. M., & Clarkson, J. (2009). Health care process modelling: Which method when? *International Journal for Quality in Health Care*, *21*(3), 214–224.

Kannampallil, T. G., Goss, C. W., Evanoff, B. A., Strickland, J. R., McAlister, R. P., & Duncan, J. (2020). Exposure to COVID-19 patients increases physician trainee stress and burnout. *PloS one*, *15*(8), e0237301.

Karasek, R. A. (1979). Job demands, job decision latitude, and mental strain: Implications for job redesign. *Administrative Science Quarterly*, *24*, 285–308.

Karasek, R. A., & Theorell, T. (1990). *Healthy Work: Stress, Productivity and the Reconstruction of Working Life*. Basic Books.

Karsh, B.-T. (2004). Beyond usability: Designing effective technology implementation systems to promote patient safety. *Quality and Safety in Health Care*, *13*, 388–394.

Karsh, B.-T., & Brown, R. (2010). Macroergonomics and patient safety: The impact of levels on theory, measurement, analysis and intervention in patient safety research. *Applied Ergonomics*, *41*(5), 674–681.

Karsh, B.-T., Carayon, P., Smith, M., Skibinski, K., Thomadsen, B., Brennan, P., & Murray, M. E. (2005). The University of Wisconsin-Madison multidisciplinary graduate certificate in patient safety. In K. Henriksen, J. B. Battles, E. E. Marks, & D. I. Lewin (Eds.), *Advances in Patient Safety: From Research to Implementation* (Vol. 4, pp. 269–281). Agency for Healthcare Research and Quality.

Karsh, B.-T., Holden, R. J., Alper, S. J., & Or, C. K. L. (2006). A human factors engineering paradigm for patient safety: Designing to support the performance of the healthcare professional. *Quality & Safety in Health Care*, *15*(i6), i59–i65.

Karsh, B.-T., Waterson, P., & Holden, R. J. (2014). Crossing levels in systems ergonomics: A framework to support 'mesoergonomic' inquiry. *Applied Ergonomics*, *45*(1), 45–54.

Katz, D., & Kahn, R. L. (1966). *The Social Psychology of Organizations*. Wiley.

Kaushal, R., Bates, D. W., Landrigan, C., & et al. (2001). Medication errors and adverse drug events in pediatric inpatients. *JAMA*, *285*(16), 2114–2120.

Kebapci, A., & Ozkaynak, M. (2022). Endotracheal tube cuff pressure management: An observational study guided by the SEIPS model. *Dimensions of Critical Care Nursing*, *41*(2), 64–75.

Keenan, J. E., Speicher, P. J., Thacker, J. K. M., Walter, M., Kuchibhatla, M., & Mantyh, C. R. (2014). The preventive surgical site infection bundle in colorectal surgery: An effective approach to surgical site infection reduction and health care cost savings. *JAMA Surgery*, *149*(10), 1045–1052.

Kelly, M. M., Xie, A., Carayon, P., DuBenske, L. L., Ehlenbach, M. L., & Cox, E. D. (2013). Strategies for improving family engagement during family-centered rounds. *Journal of Hospital Medicine*, *8*(4), 201–207.

Kelly, M. M., Xie, A., Li, Y., Cartmill, R., Cox, E. D., Brown, R. L., Wetterneck, T., & Carayon, P. (2019). System factors influencing the use of a family-centered rounds checklist. *Pediatric Quality & Safety*, *4*(4), 1–7.

King, H. B., Battles, J., Baker, D. P., Alonso, A., Salas, E., Webster, J., Toomey, L., & Salisbury, M. (2008). TeamSTEPPS™: Team strategies and tools to enhance performance and patient safety. In K. Henriksen, J. B. Battles, & M. A. Keyes (Eds.), *Advances in Patient Safety: New Directions and Alternative Approaches* (Vol. 3). Agency for Healthcare Research and Quality.

Kirkendall, E. S., Brady, P. W., Corathers, S. D., Ruddy, R. M., Fox, C., Nelson, H., Wetterneck, T. B., Rodgers, I., & Walsh, K. E. (2023). Safer type 1 diabetes care at home: SEIPS-based process mapping with parents and clinicians. *Pediatric Quality & Safety*, *8*(3), e649.

Kirwan, B., & Ainsworth, L. K. (1992). *A Guide to Task Analysis*. Taylor & Francis Group.

Klein, L. (2014). What do we actually mean by 'sociotechnical'? On values, boundaries and the problems of language. *Applied Ergonomics*, *45*(2, Part A), 137–142.

Kleiner, B. M. (2006). Macroergonomics: Analysis and design of work systems. *Applied Ergonomics*, *37*(1), 81–89.

Kleiner, B. M. (2007). Sociotechnical system design in health care. In P. Carayon (Ed.), *Handbook of Human Factors and Ergonomics in Health Care and Patient Safety* (pp. 79–94). Lawrence Erlbaum Associates.

Kleiner, B. M. (2008). Macroegonomics: Work system analysis and design. *Human Factors*, *50*(3), 461–467.

Knutsen Glette, M., Ludlow, K., Wiig, S., Bates, D. W., & Austin, E. E. (2023). Resilience perspective on healthcare professionals' adaptations to changes and challenges resulting from the COVID-19 pandemic: A meta-synthesis. *BMJ Open*, *13*(9), e071828.

Kohn, L. T., Corrigan, J. M., & Donaldson, M. S. (2000). *To Err is Human: Building a Safer Health System*. The National Academies Press.

Komashie, A., Ward, J., Bashford, T., Dickerson, T., Kaya, G. K., Liu, Y., Kuhn, I., Günay, A., Kohler, K., Boddy, N., O'Kelly, E., Masters, J., Dean, J., Meads, C., & Clarkson, P. J. (2021). Systems approach to health service design, delivery and improvement: A systematic review and meta-analysis. *BMJ Open*, *11*(1), e037667.

Korunka, C., Weiss, A., & Karetta, B. (1993). Effects of new technologies with special regard for the implementation process per se. *Journal of Organizational Behavior*, *14*(4), 331–348.

Krauss, M. J., Evanoff, B., Hitcho, E., Ngugi, K. E., Dunagan, W. C., Fischer, I., Birge, S., Johnson, S., Costantinou, E., & Fraser, V. J. (2005). A case-control study of patient, medication, and care-related risk factors for inpatient falls. *Journal of General Internal Medicine*, *20*(2), 116–122.

Kuitunen, S., Saksa, M., & Holmström, A.-R. (2024). Medication errors and error chains involving high-alert medications in a paediatric hospital setting: A qualitative analysis of self-reported medication safety incidents. *Drugs - Real World Outcomes*, *12*(1), 45–61.

Kushniruk, A., & Nohr, C. (2016). Participatory design, user involvement and health IT evaluation. In E. Ammenwerth & M. Rigby (Eds.), *Evidence-Based Health Informatics* (pp. 139–151). IOS Press.

Kwan, W.-M., Mok, C.-K., Kwok, Y.-T., Ling, H., Lam, H.-W., Law, T.-H., Leung, P.-M., Mak, M.-Y., Que, T.-L., Kan, C.-H., & Tang, Y.-H. (2021). Using the Systems Engineering Initiative for Patient Safety (SEIPS) model to describe the planning and management of the COVID-19 pandemic in Hong Kong. *International Journal for Quality in Health Care*, *33*(1), mzab042.

Kwon, J. H., Advani, S. D., Branch-Elliman, W., Braun, B. I., Cheng, V. C.-C., Chiotos, K., Douglas, P., Gohil, S. K., Keller, S. C., Klein, E. Y., Krein, S. L., Lofgren, E. T., Merrill, K., Moehring, R. W., Monsees, E., Perri, L., Scaggs Huang, F., Shelly, M. A., Skelton, F., … Mody, L. (2024). A call to action: The SHEA research agenda to combat healthcare-associated infections. *Infection Control & Hospital Epidemiology*, *45*(9), 1023–1040.

Ladell, M. M., Jacobson, N. L., Yale, S. C., McDermott, K. L., Papautsky, E. L., Catchpole, K. R., & Scanlon, M. C. (2025). The problem with how we view medical (and diagnostic) error in emergency medicine. *Academic Emergency Medicine*, *32*(3), 340–347.

Lamé, G., Liberati, E. G., Canham, A., Burt, J., Hinton, L., Draycott, T., Winter, C., Dakin, F. H., Richards, N., Miller, L., Willars, J., & Dixon-Woods, M. (2024). Why is safety in intrapartum electronic fetal monitoring so hard? A qualitative study combining human factors/ergonomics and social science analysis. *BMJ Quality & Safety*, *33*, 246–256.

Landa-Avila, I. C., Escobar-Tello, C., Jun, G. T., & Cain, R. (2021). Multiple outcome interactions in healthcare systems: A participatory outcome mapping approach. *Ergonomics*, *65*(3), 362–383.

Lanham, H. J., Leykum, L. K., Taylor, B. S., McCannon, C. J., Lindberg, C., & Lester, R. T. (2013). How complexity science can inform scale-up and spread in health care: Understanding the role of self-organization in variation across local contexts. *Social Science & Medicine*, *93*, 194–202.

Larsen, E. P., Schaeubinger, M. M., Won, J., Sze, R. W., & Anupindi, S. (2024). Integrating human factors engineering into your pediatric radiology practice. *Pediatric Radiology*, *54*(6), 936–943.

Larson, C. M., & Saine, D. (2013). *Medication Safety Officer's Handbook*. American Society of Health-System Pharmacists.

Lawton, R. (2018). It ain't what you do (but the way that you do it): Will safety II transform the way we do patient safety? Comment on "False dawns and new horizons in patient safety research and practice". *International Journal of Health Policy and Management*, *7*(7), 659–661.

Le Coze, J. C. (2022). The 'new view' of human error: Origins, ambiguities, successes and critiques. *Safety Science*, *154*.

Le Coze, J.-C., & Reiman, T. (Eds.). (2023). *Visualizing Safety, An Exploration - Drawings, Pictures, Images, Videos and Images*. Springer.

Leon, C., Hogan, H., & Jani, Y. H. (2024). Seeking systems-based facilitators of safety and healthcare resilience: A thematic review of incident reports. *International Journal for Quality in Health Care*, *36*(3), mzae057.

Leplat, J. (1989). Error analysis, instrument and object of task analysis. *Ergonomics*, *32*(7), 813–822.

Leveson, N. (2020). *Safety III: A Systems Approach to Safety and Resilience*. MIT Engineering Systems Lab.

Leveson, N., Samost, A., Dekker, S., Finkelstein, S., & Raman, J. (2020). A systems approach to analyzing and preventing hospital adverse events. *Journal of Patient Safety*, *16*(2), 162–167.

Levine, D. M., Linder, J. A., & Landon, B. E. (2016). The quality of outpatient care delivered to adults in the United States, 2002 to 2013. *JAMA Internal Medicine*, *176*(12), 1778–1790.

Levitt, P. (2014a). Challenging the systems approach: Why adverse event rates are not improving. *BMJ Quality & Safety*, *23*(12), 1051–1052.

Levitt, P. (2014b, March 16, 2014). When medical errors kills. *Los Angeles Times*.

Li, L. Z., Yang, P., Singer, S. J., Pfeffer, J., Mathur, M. B., & Shanafelt, T. (2024). Nurse burnout and patient safety, satisfaction, and quality of care: A systematic review and meta-analysis. *JAMA Network Open*, *7*(11), e2443059.

Li, Y., Carayon, P., Hundt, A. S., Kianfar, S., Hoonakker, P., Salek, D., & Tomcavage, J. (2014). Levels of strategies used by care managers to deal with health IT-related barriers. In O. Broberg, N. Fallentin, P. Hasle, P. L. Jensen, A. Kabel, M. E. Larsen, & T. Weller (Eds.), *Human Factors in Organizational Design and Management - XI* (pp. 469–473). IEA Press.

Lim, S. Y., & Murphy, L. R. (1999). The relationship of organizational factors to employee health and overall effectiveness. *American Journal of Industrial Medicine, supplement 1*, 64–65.

Lindblom, C. E. (1959). The science of "muddling through". *Public Administration Review*, *19*(2), 79–88.

Linfors, E. W., & Neelon, F. A. (1980). The case for bedside rounds. *New England Journal of Medicine*, *303*(21), 1230–1233.

Lintern, G., & Kugler, P. N. (2022). Emergence and non-emergence for system safety. *Theoretical Issues in Ergonomics Science*, *24*(6), 646–661.

Lipitz-Snyderman, A., Steinwachs, D., Needham, D. M., Colantuoni, E., Morlock, L. L., & Pronovost, P. J. (2011). Impact of a statewide intensive care unit quality improvement initiative on hospital mortality and length of stay: Retrospective comparative analysis. *BMJ*, *342*, d219.

Loeppke, R., Boldrighini, J., Bowe, J., Braun, B., Eggins, E., Eisenberg, B. S., Grundy, P., Hohn, T., Hudson, T. W., Kannas, J., Jr., Kapp, E. A., Konicki, D., Larson, P., McCutcheon, S., McLellan, R. K., Ording, J., Perkins, C., Russi, M., Stutts, C., & Yarbrough, M. (2017). Interaction of health care worker health and safety and patient health and safety in the US health care system: Recommendations from the 2016 summit. *Journal of Occupational and Environmental Medicine*, *59*(8), 803–813.

Long, S., & Vincent, C. (2012). Patient safety. In S. J. Sinclair, J. E. Morley, & B. Vellas (Eds.), *Principles and Practice of Geriatric Medicine* (pp. 153–170). John Wiley & Sons, Ltd.

Look, K. A., & Stone, J. A. (2018). Medication management activities performed by informal caregivers of older adults. *Research in Social and Administrative Pharmacy*, *14*(5), 418–426.

Lundstrom, T., Pugliese, G., Bartley, J., Cox, J., & Guither, C. (2002). Organizational and environmental factors that affect worker health and safety and patient outcomes. *American Journal of Infection Control*, *30*(2), 93–106.

Maben, J., Ball, J., & Edmondson, A. C. (2023). *Workplace Conditions*. Cambridge University Press.

Macrae, C. (2014). Early warnings, weak signals and learning from healthcare disasters. *BMJ Quality & Safety*, *23*, 440–445.

Magerøy, M. R., Braut, G. S., Macrae, C., & Wiig, S. (2023). Healthcare leaders' and elected politicians' approach to support-systems and requirements for complying with quality and safety regulation in nursing homes: A case study. *BMC Health Services Research*, *23*, 1–15.

Magerøy, M. R., Macrae, C., Braut, G. S., & Wiig, S. (2024). Managing patient safety and staff safety in nursing homes: Exploring how leaders of nursing homes negotiate their dual responsibilities-a case study. *Frontiers in Health Services*, *4*, 1275743.

Magill, S. S., Edwards, J. R., Bamberg, W., Beldavs, Z. G., Dumyati, G., Kainer, M. A., Lynfield, R., Maloney, M., McAllister-Hollod, L., Nadle, J., Ray, S. M., Thompson, D. L., Wilson, L. E., & Fridkin, S. K. (2014). Multistate point-prevalence survey of health care–associated infections. *New England Journal of Medicine*, *370*(13), 1198–1208.

Mannion, R., & Braithwaite, J. (2017). False dawns and new horizons in patient safety research and practice. *International Journal of Health Policy and Management*, *6*(12), 685–689.

Marcilly, R., Zheng, W. Y., Quindroit, P., Pelayo, S., Berdot, S., Charpiat, B., Corny, J., Drouot, S., Frery, P., Leguelinel-Blache, G., Mondet, L., Potier, A., Robert, L., Ferret, L., & Baysari, M. (2023). Comparison of the validity, perceived usefulness, and usability of I-MeDeSA and TEMAS, two tools to evaluate alert system usability. *International Journal of Medical Informatics*, *175*, 105091.

Markopoulos, D., Tsolakidis, A., Karanikolas, N., Marinagi, A., & Skourlas, C. (2023). *Applying Soft System Methodology for a clearer understanding of the future Intensive Care Units Proceedings of the 27th Pan-Hellenic Conference on Progress in Computing and Informatics*. Lamia, Greece.

Martin, G., Stanford, S., & Dixon-Woods, M. (2023). A decade after Francis: Is the NHS safer and more open? *BMJ*, *380*, 1–2.

Maslach, C., Schaufeli, W. B., & Leiter, M. P. (2001). Job burnout. *Annual Review of Psychology*, *52*, 397–422.

Masterson, D., Areskoug Josefsson, K., Robert, G., Nylander, E., & Kjellström, S. (2022). Mapping definitions of co-production and co-design in health and social care: A systematic scoping review providing lessons for the future. *Health Expectations*, *25*(3), 902–913.

Mata, D. A., Ramos, M. A., Bansal, N., Khan, R., Guille, C., Di Angelantonio, E., & Sen, S. (2015). Prevalence of depression and depressive symptoms among resident physicians: A systematic review and meta-analysis. *JAMA*, *314*(22), 2373–2383.

Mazor, K. M., Goff, S. L., Dodd, K. S., Velten, S. J., & Walsh, K. E. (2010). Parents' perceptions of medical errors. *Journal of Patient Safety*, *6*(2), 102–107.

McCaughey, D., Kimmel, A., Savage, G., Lukas, T., Walsh, E., & Halbesleben, J. (2016). Antecedents to workplace injury in the health care industry: A synthesis of the literature. *Health Care Management Review*, *41*(1), 42–55.

McCulloch, P. (2024). Fixing patient safety: Are we nearly there yet? *BMJ Quality & Safety*, *33*(8), 539–542.

McGlynn, E. A., Asch, S. M., Adams, J., Keesey, J., Hicks, J., DeCristofaro, A., & Kerr, E. (2003). The quality of health care delivered to adults in the United States. *The New England Journal of Medicine*, *348*(26), 2635–2645.

McLean, K. J., Haas, M., Koenig, J., Horvath, M., Vigil, M., Werner, N. E., & Bishop, L. (2024). "I'm dealing with a health care system that doesn't get it": Barriers and facilitators to inclusive healthcare for autistic adults. *Autism*, *28*(6), 1382–1393.

McNab, D., McKay, J., Shorrock, S., Luty, S., & Bowie, P. (2020). Development and application of 'systems thinking' principles for quality improvement. *BMJ Open Quality*, *9*(1), e000714.

McTiernan, P., Wachter, R. M., Meyer, G. S., & Gandhi, T. K. (2015). Patient safety is not elective: A debate at the NPSF Patient Safety Congress. *BMJ Quality & Safety*, *24*(2), 162–166.

Meadows, D. H. (2008). *Thinking in Systems: A Primer*. Chelsea Green Publishing.

Meister, D., & Enderwick, T. P. (2001). *Human Factors in System Design, Development, and Testing*. Lawrence Erlbaum Associates.

Melles, M., Albayrak, A., & Goossens, R. (2021). Innovating health care: Key characteristics of human-centered design. *International Journal for Quality in Health Care, 33*(Supplement_1), 37–44.

Melnick, E. R., Dyrbye, L. N., Sinsky, C. A., Trockel, M., West, C. P., Nedelec, L., Tutty, M. A., & Shanafelt, T. (2019). The association between perceived electronic health record usability and professional burnout among US physicians. *Mayo Clinic Proceedings, 95*(3), 476–487.

Merton, R. K. (1936). The unanticipated consequences of purposive social action. *American Sociological Review, 1*(6), 894–904.

Meshkati, N., Tabibzadeh, M., Farshid, A., Rahimi, M., & Alhanaee, G. (2016). People-technology-ecosystem integration: A framework to ensure regional interoperability for safety, sustainability, and resilience of interdependent energy, water, and seafood sources in the (Persian) Gulf. *Human Factors, 58*(1), 43–57.

Mid Staffordshire NHS Foundation Trust Public Inquiry. (2013). *Report of the Mid Staffordshire NHS Foundation Trust public inquiry*. Stationery Office.

Mitchell, P., Wynia, M., Golden, R., McNellis, B., Okun, S., Webb, E. C., Rohrbach, V., & Von Kohorn, I. (2012). *Core Principles & Values of Effective Team-Based Health Care*. Institute of Medicine.

Mollo, V., & Falzon, P. (2004). Auto- and allo-confrontation as tools for reflective activities. *Applied Ergonomics, 35*(6), 531–540.

Mollo, V., & Falzon, P. (2008). The development of collective reliability: A study of therapeutic decision-making. *Theoretical Issues in Ergonomics Science, 9*(3), 223–254.

Monat, J. P., & Ganon, T. F. (2015). What is systems thinking? A review of selected literature plus recommendations. *American Journal of Systems Science, 4*(1), 11–26.

Moon, S. E. J., Hogden, A., & Eljiz, K. (2022). Sustaining improvement of hospital-wide initiative for patient safety and quality: A systematic scoping review. *BMJ Open Quality, 11*(4), 1–12.

Moray, N. (2000). Culture, politics and ergonomics. *Ergonomics, 43*(7), 858–868.

Mørch, A. I., & Mehandjiev, N. D. (2000). Tailoring as collaboration: The mediating role of multiple representations and application units. *Computer Supported Cooperative Work, 9*(1), 75–100.

Morey, J. C., Simon, R., Jay, G. D., Wears, R. L., Salisbury, M., Dukes, K. A., & Berns, S. D. (2002). Error reduction and performance improvement in the emergency department through formal teamwork training: Evaluation results of the MedTeams project. *Health Services Research, 37*(6), 1553–1581.

Morimoto, T., Sakuma, M., Matsui, K., Kuramoto, N., Toshiro, J., Murakami, J., Fukui, T., Saito, M., Hiraide, A., & Bates, D. W. (2011). Incidence of adverse drug events and medication errors in Japan: The JADE study. *Journal of General Internal Medicine, 26*(2), 148–153.

Muething, S. E., Kotagal, U. R., Schoettker, P. J., del Rey, J. G., & DeWitt, T. G. (2007). Family-centered bedside rounds: A new approach to patient care and teaching. *Pediatrics, 119*(4), 829–832.

Mumford, E. (2003). *Redesigning Human Systems*. IRM Press.

Musuuza, J. S., Hundt, A. S., Carayon, P., Christensen, K., Ngam, C., Haun, N., & Safdar, N. (2019). Implementation of a *Clostridioides difficile* prevention bundle: Understanding common, unique, and conflicting work system barriers and facilitators for subprocess design. *Infection Control & Hospital Epidemiology, 40*(8), 880–888.

Nascimento, A., & Falzon, P. (2012). Producing effective treatment, enhancing safety: Medical physicists' strategies to ensure quality in radiotherapy. *Applied Ergonomics, 43*(4), 777–784.

NASEM. (2018). *Crossing the Global Quality Chasm*. The National Academies Press.

NASEM. (2019). *Taking Action Against Clinician Burnout: A Systems Approach to Professional Well-Being*. The National Academies Press.
National Academy of Medicine (NAM). (2025). *Generative Artificial Intelligence in Health and Medicine: Opportunities and Responsibilities for Transformative Innovation*. The National Academies Press.
Ngam, C., Schoofs Hundt, A., Haun, N., Carayon, P., Stevens, L., & Safdar, N. (2017). Barriers and facilitators to Clostridium difficile infection prevention: A nursing perspective. *American Journal of Infection Control*, *45*(12), 1363–1368.
Nilsen, P. (2024). The historical background of implementation science. In P. Nilsen (Ed.), *Implementation Science - Theory and Applications* (pp. 12–21). Routledge.
Norman, D. (2023). *Design for a Better World*. MIT Press.
Norman, D. A., & Draper, S. W. (1986). *User-Centered System Design: New Perspectives on Human-Computer Interaction*. Lawrence Erlbaum Associates.
Norman, D. A., & Stappers, P. J. (2015). DesignX: Complex sociotechnical systems. *She Ji: The Journal of Design, Economics, and Innovation*, *1*(2), 83–106.
Noro, K., & Imada, A. (1991). *Participatory Ergonomics*. Taylor & Francis Group.
Nundy, S., Cooper, L. A., & Mate, K. S. (2022). The Quintuple Aim for health care improvement: A new Imperative to advance health equity. *JAMA*, *327*(6), 521–522.
Nyssen, A.-S., & Blavier, A. (2006). Error detection: A study in anaesthesia. *Ergonomics*, *49*(5–6), 517–525.
Obradovich, J. H., & Woods, D. D. (1996). Users as designers: How people cope with poor HCI design in computer-based medical devices. *Human Factors*, *38*(4), 574–592.
Odberg, K. R., Aase, K., Grusd, E., & Vifladt, A. (2025). The work system of prehospital medication administration: A qualitative mixed methods study with ambulance professionals. *BMC Emergency Medicine*, *25*(1), 54.
Odberg, K. R., Hansen, B. S., Aase, K., & Wangensteen, S. (2020). A work system analysis of the medication administration process in a Norwegian nursing home ward. *Applied Ergonomics*, *86*, 103100.
OECD. (2024). *The Economics of Diagnostic Safety*. OECD.
Oh, S., Sang, E., Stawnychy, M. A., Garren, P., You, S. B., O'Connor, M., Hirschman, K. B., Hodgson, N., Cranston, T., Jablonski, J., O'Brien, K., Newcomb, M., Spahr, M., & Bowles, K. H. (2024). Application of a Human Factors and Systems Engineering approach to explore care transitions of sepsis survivors from hospital to home health care. *Human Factors*, *66*(11), 2468–2484.
O'Hara, J. K., Baxter, R., & Hardicre, N. (2020). 'Handing over to the patient': A FRAM analysis of transitional care combining multiple stakeholder perspectives. *Applied Ergonomics*, *85*, 103060.
Olds, D. M., & Clarke, S. P. (2010). The effect of work hours on adverse events and errors in health care. *Journal of Safety Research*, *41*(2), 153–162.
OSHA. (2024). *Integrating Patient and Workplace Safety Programs*. Retrieved October 24, 2024 from https://www.osha.gov/sites/default/files/publications/OSHA3730.pdf
Oshry, B. (2007). *Seeing Systems - Unlocking the Mysteries of Organizational Life*. Berrett-Koehler Publishers.
Østergaard, K. L., Simonsen, J., & Karasti, H. (2018). Examining situated design practices: Nurses' transformations towards genuine participation. *Design Studies*, *59*, 37–57.
Page, B., Irving, D., Amalberti, R., & Vincent, C. (2024). Health services under pressure: A scoping review and development of a taxonomy of adaptive strategies. *BMJ Quality & Safety*, *33*, 738–747.
Page, B., Irving, D., Carthey, J., Welch, J., Higham, H., & Vincent, C. (2025). Strategies for adapting under pressure: An interview study in intensive care units. *BMJ Quality & Safety*, *34*(2), 81–91.

Papazian, L., Klompas, M., & Luyt, C. E. (2020). Ventilator-associated pneumonia in adults: A narrative review. *Intensive Care Medicine*, *46*(5), 888–906.

Parand, A., Benn, J., Burnett, S., Pinto, A., & Vincent, C. (2012). Strategies for sustaining a quality improvement collaborative and its patient safety gains. *International Journal for Quality in Health Care*, *24*(4), 380–390.

Parasuraman, R., Sheridan, T. B., & Wickens, C. D. (2000). A model for types and levels of human interaction with automation. *IEEE Transactions on Systems Man and Cybernetics Part a-Systems and Humans*, *30*(3), 286–297.

Parker, S. K. (2014). Beyond motivation: Job and work design for development, health, ambidexterity, and more. *Annual Review of Psychology*, *65*, 661–691.

Parker, S. K., Wall, T. D., & Cordery, J. L. (2001). Future work design research and practice: Towards an elaborated model of work design. *Journal of Occupational & Organizational Psychology*, *74*(4), 413.

Parvin, N., & Pollock, A. (2020). Unintended by design: On the political uses of "Unintended Consequences". *Engaging Science, Technology, and Society*, *6*, 320–327.

Pasmore, W. A. (1988). *Designing Effective Organizations: The Sociotechnical Systems Perspective*. John Wiley & Sons.

Payne, R., Clarke, A., Swann, N., van Dael, J., Brenman, N., Rosen, R., Mackridge, A., Moore, L., Kalin, A., Ladds, E., Hemmings, N., Rybczynska-Bunt, S., Faulkner, S., Hanson, I., Spitters, S., Wieringa, S., Dakin, F. H., Shaw, S. E., Wherton, J., … Greenhalgh, T. (2024). Patient safety in remote primary care encounters: Multimethod qualitative study combining Safety I and Safety II analysis. *BMJ Quality & Safety*, *33*, 573–586.

Perrow, C. (1983). The organizational context of human factors engineering. *Administrative Science Quarterly*, *28*(4), 521–541.

Perry, S. J., Catchpole, K., Rivera, A. J., Henrickson Parker, S., & Gosbee, J. (2021). 'Strangers in a strange land': Understanding professional challenges for human factors/ergonomics and healthcare. *Applied Ergonomics*, *94*, 103040.

Perry, S. J., Wears, R. L., & Fairbanks, R. J. (2012). Human factors and ergonomics in the emergency department. In P. Carayon (Ed.), *Handbook of Human Factors and Ergonomics in Health Care and Patient Safety* (Second ed., pp. 709–722). Taylor & Francis Group.

Peryer, G., Kelly, S., Blake, J., Burton, J. K., Irvine, L., Cowan, A., Akdur, G., Killett, A., Brand, S. L., Musa, M. K., Meyer, J., Gordon, A. L., & Goodman, C. (2022). Contextual factors influencing complex intervention research processes in care homes: A systematic review and framework synthesis. *Age and Ageing*, *51*(3), 1–16.

Peters, L. H., & O'Connor, E. J. (1980). Situational constraints and work outcomes: The influences of a frequently overlooked construct. *Academy of Management Review*, *5*(3), 391–397.

Pickup, L., Atkinson, S., Hollnagel, E., Bowie, P., Gray, S., Rawlinson, S., & Forrester, K. (2017). Blood sampling: Two sides to the story. *Applied Ergonomics*, *59*(Pt A), 234–242.

Pickup, L., Nugent, B., & Bowie, P. (2019). A preliminary ergonomic analysis of the MRI work system environment: Implications and recommendations for safety and design. *Radiography*, *25*(4), 339–345.

Plsek, P., & Greenhalgh, T. (2001). Complexity science: The challenge of complexity in health care. *BMJ*, *323*, 625–628.

Podtschaske, B., Salazar, N., & Rao, M. B. (2018). Proactive, transdisciplinary safer-system redesign. A field report from the hospital frontline. *IISE Transactions on Occupational Ergonomics and Human Factors*, *6*, 192–199.

Ponnala, S., Block, L., Lingg, A. J., Kind, A. J., & Werner, N. E. (2020). Conceptualizing caregiving activities for persons with dementia (PwD) through a patient work lens. *Applied Ergonomics*, *85*, 103070.

President's Council of Advisors on Science and Technology. (2023). *A Transformational Effort on Patient Safety*. Executive Office of the President.

Proctor, R. W., McLaughlin, A. C., Szalma, J. L., Blickensderfer, E. L., & Vu, K. L. (2021). Division 21 has been devoted to human-centered design since the 1950s: Comment on Lyon, Brewer, and Arean (2020). *American Psychologist*, *76*(7), 1186–1188.

Pronovost, P., Needham, D., Berenholtz, S., Sinopoli, D., Chu, H., Cosgrove, S., Sexton, B., Hyzy, R., Welsh, R., Roth, G., Bander, J., Kepros, J., & Goeschel, C. (2006). An intervention to decrease catheter-related bloodstream infections in the ICU. *New England Journal of Medicine*, *355*(26), 2725–2732.

Pronovost, P. J., & Bo-Linn, G. W. (2012). Preventing patient harms through systems of care. *JAMA*, *308*(8), 769–770.

Pronovost, P. J., Goeschel, C. A., Colantuoni, E., Watson, S., Lubomski, L. H., Berenholtz, S. M., Thompson, D. A., Sinopoli, D. J., Cosgrove, S., Sexton, J. B., Marsteller, J. A., Hyzy, R. C., Welsh, R., Posa, P., Schumacher, K., & Needham, D. (2010). Sustaining reductions in catheter related bloodstream infections in Michigan intensive care units: Observational study. *BMJ*, *340*(feb04_1), c309.

Pronovost, P. J., Sutcliffe, K. M., Basu, L., & Dixon-Woods, M. (2017). Changing the narratives for patient safety. *Bulletin of the World Health Organization*, *95*(6), 478–480.

Randhawa, G. S., Xiao, Y., & Gorman, P. N. (2019). Designing a "Thinking System" to reduce the human burden of care delivery. *eGEMS*, *7*(1), 1–4.

Rasmussen, J. (2000). Human factors in a dynamic information society: Where are we heading? *Ergonomics*, *43*(7), 869–879.

Rasooly, I. R., Marshall, T. L., Cifra, C. L., Catchpole, K., Kuzma, N. C., Brady, P. W., Melton, K., Khan, A., Chien, A. T., Lipstein, E. A., Landrigan, C. P., & Walsh, K. E. (2024). Developing methods to identify resilience and improve communication about diagnosis in pediatric primary care. *Frontiers in Medicine*, *11*, 1414892.

Ratwani, R. M., Savage, E., Will, A., Fong, A., Karavite, D., Muthu, N., Rivera, A. J., Gibson, C., Asmonga, D., Moscovitch, B., Grundmeier, R., & Rising, J. (2018). Identifying electronic health record usability and safety challenges in pediatric settings. *Health Affairs*, *37*(11), 1752–1759.

Read, G. J. M., Shorrock, S., Walker, G. H., & Salmon, P. M. (2021). State of science: evolving perspectives on 'human error'. *Ergonomics*, *64*(9), 1091–1114.

Reason, J. (1997). *Managing the Risks of Organizational Accidents*. Ashgate.

Reay, S. D., Collier, G., Douglas, R., Hayes, N., Nakarada-Kordic, I., Nair, A., & Kennedy-Good, J. (2017). Prototyping collaborative relationships between design and healthcare experts: Mapping the patient journey. *Design for Health*, *1*(1), 65–79.

Reid, P. R., Compton, W. D., Grossman, J. H., & Fanjiang, G. (2005). *Building a Better Delivery System. A New Engineering/Health Care Partnership*. The National Academies Press.

Rello, J., Chastre, J., Cornaglia, G., & Masterton, R. (2011). A European care bundle for management of ventilator-associated pneumonia. *Journal of Critical Care*, *26*(1), 3–10.

Reme, S. E., Dennerlein, J. T., Hashimoto, D., & Sorensen, G. (2012). Musculoskeletal pain and psychological distress in hospital patient care workers. *Journal of Occupational Rehabilitation*, *22*(4), 503–510.

Reme, S. E., Shaw, W. S., Boden, L. I., Tveito, T. H., O'Day, E. T., Dennerlein, J. T., & Sorensen, G. (2014). Worker assessments of organizational practices and psychosocial work environment are associated with musculoskeletal injuries in hospital patient care workers. *American Journal of Industrial Medicine*, *57*(7), 810–818.

Robert, G., Locock, L., Williams, O., Cornwell, J., Donetto, S., & Goodrich, J. (2022). *Co-Producing and Co-Designing*. Cambridge University Press.

Rodriguez, N. M., Burleson, G., Linnes, J. C., & Sienko, K. H. (2023). Thinking beyond the device: An overview of human- and equity-centered approaches for health technology design. *Annual Review of Biomedical Engineering*, *25*, 257–280.

Rodriguez, R. M., Medak, A. J., Baumann, B. M., Lim, S., Chinnock, B., Frazier, R., & Cooper, R. J. (2020). Academic emergency medicine physicians' anxiety levels, stressors, and potential stress mitigation measures during the acceleration phase of the COVID-19 pandemic. *Academic Emergency Medicine*, *27*(8), 700–707.

Rogers, A. E. (2012). Health-care work schedules. In P. Carayon (Ed.), *Handbook of Human Factors and Ergonomics in Health Care and Patient Safety* (Second ed., pp. 199–246). Taylor & Francis Group.

Rouse, W. B. (2008). Health care as a complex adaptive system: Implications for design and management. *The Bridge*, *38*(1), 17.

Rousseau, D., & House, R. J. (1994). Meso organizational behavior: Avoiding three fundamental biases. *Journal of Organizational Behavior*, *13*.

Rousseau, D. M., & Fried, Y. (2001). Location, location, location: Contextualizing organizational research. *Journal of Organizational Behavior*, *22*(1), 1–13.

Russ, A. L., Fairbanks, R. J., Karsh, B. T., Militello, L. G., Saleem, J. J., & Wears, R. L. (2013). The science of human factors: Separating fact from fiction. *BMJ Quality & Safety*, *22*(10), 802–808.

Rutkowski, R. A., Carayon, P., Hoonakker, P., Pulia, M. S., Shah, M. N., Patterson, B. W., & Werner, N. E. (2025). (con)Figuring out influence: A modified Delphi approach to configural diagramming to identify influential work system factors on emergency department disposition decision-making. *Submitted.*

Safren, M. A., & Chapanis, A. (1960a). A critical incident study of hospital medication errors - Part 1. *Hospitals*, *34*, 32–34, 57–66.

Safren, M. A., & Chapanis, A. (1960b). A critical incident study of hospital medication errors - Part 2. *Hospitals*, *34*, 53, 65–68.

Sainfort, F., Karsh, B., Booske, B. C., & Smith, M. J. (2001). Applying quality improvement principles to achieve healthy work organizations. *Journal on Quality Improvement*, *27*(9), 469–483.

Sakuma, M., Ohta, Y., Takeuchi, J., Yuza, Y., Ida, H., Bates, D. W., & Morimoto, T. (2024). Adverse events in pediatric patients: The Japan Adverse Event study. *Journal of Patient Safety*, *20*, 38–44.

Salahuddin, L., & Ismail, Z. (2015). Classification of antecedents towards safety use of health information technology: A systematic review. *International Journal of Medical Informatics*, *11*, 877–891.

Salas, E., & Rosen, M. A. (2013). Building high reliability teams: Progress and some reflections on teamwork training. *BMJ Quality & Safety*, *22*(5), 369–373.

Salas, E., Wilson, K. A., Murphy, C. E., King, H., & Salisbury, M. (2008). Communicating, coordinating, and cooperating - When lives depend on it: Tips for teamwork. *The Joint Commission Journal on Quality and Patient Safety*, *34*(6), 333–341.

Salvendy, G., & Karwoski, W. (Eds.). (2021). *Handbook of Human Factors and Ergonomics* (Fifth ed.). Wiley.

Salwei, M., & Carayon, P. (2022). A sociotechnical systems framework for the application of artificial intelligence in health care delivery. *Journal of Cognitive Engineering and Decision Making*, *16*(4), 194–206.

Salwei, M. E., Carayon, P., Hundt, A. S., Hoonakker, P., Agrawal, V., Kleinschmidt, P., Stamm, J., Wiegmann, D., & Patterson, B. W. (2019). Role network measures to assess health-care team adaptation to complex situations: The case of venous thromboembolism prophylaxis. *Ergonomics*, *62*(7), 864–879.

Sánchez de la Guía, L., Puyuelo Cazorla, M., & de-Miguel-Molina, B. (2017). Terms and meanings of "participation" in product design: From "user involvement" to "co-design". *The Design Journal*, *20*(sup1), S4539–S4551.

Sanders, L., & Stappers, P. J. (2013). *Convivial Toolbox: Generative Research for the Front End of Design*. BIS Publishers.

Sanders, L., & Stappers, P. J. (2014). From designing to co-designing to collective dreaming: three slices in time. *Interactions*, *21*(6), 24–33.

Santos, R. (2020). Reengineer healthcare: A human factors and ergonomics framework to improve the socio-technical system. *International Journal for Quality in Health Care*, *33*(Supplement_1), 19–24.

Sarkar, U., Simchowitz, B., Bonacum, D., Strull, W., Lopez, A., Rotteau, L., & Shojania, K.G. (2014). A qualitative analysis of physician perspectives on missed and delayed outpatient diagnosis: The focus on system-related factors. *The Joint Commission Journal on Quality and Patient Safety*, *40*(10), 461–470.

Sarter, N. B., Woods, D. D., & Billings, C. E. (1997). Automation surprises. In G. Salvendy (Ed.), *Handbook of Human Factors and Ergonomics* (Second ed., pp. 1926–1943). John Wiley & Sons.

Sauter, S., Lim, S. Y., & Murphy, L. R. (1996). Organizational health: A new paradigm for occupational stress at NIOSH. *Japanese Journal of Occupational Mental Health*, *4*(4), 248–254.

Scapin, D. L., & Bastien, J. M. C. (1997). Ergonomic criteria for evaluating the ergonomic quality of interactive systems. *Behaviour & Information Technology*, *16*(4), 220–231.

Scheer, E. R., Werner, N. E., Coller, R. J., Nacht, C. L., Petty, L., Tang, M., Ehlenbach, M., Kelly, M. M., Finesilver, S., Warner, G., Katz, B., Keim-Malpass, J., Lunsford, C. D., Letzkus, L., Desai, S. S., & Valdez, R. S. (2024). Designing for caregiving networks: A case study of primary caregivers of children with medical complexity. *Journal of the American Medical Informatics Association*, *31*(5), 1151–1162.

Schiff, G., & Shojania, K. G. (2022). Looking back on the history of patient safety: An opportunity to reflect and ponder future challenges. *BMJ Quality & Safety*, *31*(2), 148–152.

Schiro, J., Pelayo, S., Martinot, A., Dubos, F., Beuscart-Zéphir, M.-C., & Marcilly, R. (2020). Applying a human-centered design to develop a patient prioritization tool for a pediatric emergency department: Detailed case study of first iterations. *JMIR Human Factors*, *7*(3), e18427.

Schneider, A., Wehler, M., & Weigl, M. (2019). Effects of work conditions on provider mental well-being and quality of care: a mixed-methods intervention study in the emergency department. *BMC Emergency Medicine*, *19*(1), 1.

Schnitker, L., Martin-Khan, M., Beattie, E., & Gray, L. (2011). Negative health outcomes and adverse events in older people attending emergency departments: A systematic review. *Australasian Emergency Nursing Journal*, *14*(3), 141–162.

Scholl, I., Zill, J. M., Härter, M., & Dirmaier, J. (2014). An integrative model of patient-centeredness: A systematic review and concept analysis. *PloS one*, *9*(9), e107828.

Schroeder, M. E., Wolman, R. L., Wetterneck, T. B., & Carayon, P. (2006). Tubing misload allows free flow event with Smart intravenous infusion pump. *Anesthesiology*, *105*(2), 434–435.

Schultz, K., Carayon, P., Hundt, A. S., & Springman, S. (2007). Care transitions in the outpatient surgery preoperative process: Facilitators and obstacles to information flow and their consequences. *Cognition, Technology & Work*, *9*(4), 219–231.

Schutijser, B. C. F. M., Jongerden, I. P., Klopotowska, J. E., Portegijs, S., de Bruijne, M. C., & Wagner, C. (2019). Double checking injectable medication administration: Does the protocol fit clinical practice? *Safety Science*, *118*, 853–860.

Segarra, G. C., Catchpole, K., Rayo, M. F., Hegde, S., Jefferies, C., Woodward, J., & Taaffe, K. (2024). Revealing complex interdependencies in surgical instrument reprocessing using SEIPS 101 tools. *Applied Ergonomics*, *119*, 104307.

Seidelman, J. L., Mantyh, C. R., & Anderson, D. J. (2023). Surgical site infection prevention: A review. *JAMA*, *329*(3), 244–252.

Senders, J. W. (1994). Medical devices, medical errors, and medical accidents. In M. S. Bogner (Ed.), *Human Error in Medicine* (pp. 159–177). CRC Press.

Shah, M. N., Hollander, M. M., Jones, C. M., Caprio, T. V., Conwell, Y., Cushman, J. T., DuGoff, E. H., Kind, A. J. H., Lohmeier, M., Mi, R., & Coleman, E. A. (2018). Improving the ED-to-home transition: The community paramedic–delivered care transitions intervention: Preliminary findings. *Journal of the American Geriatrics Society, 66*(11), 2213–2220.

Shannon, E. M., Mueller, S. K., & Schnipper, J. L. (2023). Patient, caregiver, and clinician experience with a technologically enabled pillbox: A qualitative study. *ACI Open, 07*(02), e61–e70.

Shekelle, P. G., Pronovost, P., & Wachter, R. (2010). *Assessing the Evidence for Context-Sensitive Effectiveness and Safety of Patient Safety Practices: Developing Criteria.* Agency for Healthcare Research and Quality.

Shekelle, P. G., Pronovost, P. J., Wachter, R. M., McDonald, K. M., Schoelles, K., Dy, S. M., Shojania, K., Reston, J. T., Adams, A. S., Angood, P. B., Bates, D. W., Bickman, L., Carayon, P., Donaldson, L., Duan, N., Farley, D. O., Greenhalgh, T., Haughom, J. L., Lake, E., … Walshe, K. (2013). The top patient safety strategies that can be encouraged for adoption now. *Annals of Internal Medicine, 158*(5), 365–368.

Shekelle, P. G., Wachter, R., Pronovost, P., Schoelles, K., McDonald, K., Dy, S., Shojania, K., Reston, J., Berger, Z., Johnsen, B., Larkin, J., Lucas, S., Martinez, K., Motala, A., Newberry, S., Noble, M., Pfo, H. E., Ranji, S., Rennke, S., … Winters, B. (2013). *Making Health Care Safer II: An Updated Critical Analysis of the Evidence for Patient Safety Practices.* Agency for Healthcare Research and Quality.

Shojania, K., Duncan, B., McDonald, K., & Wachter, R. M. (Eds.). (2001). *Making Health Care Safer: A Critical Analysis of Patient Safety Practices.* Agency for Healthcare Research and Quality.

Shojania, K. G., & Marang-van de Mheen, P. J. (2020). Identifying adverse events: Reflections on an imperfect gold standard after 20 years of patient safety research. *BMJ Quality & Safety, 29*(4), 265–270.

Shojania, K. G., & Thomas, E. J. (2013). Trends in adverse events over time: Why are we not improving? *BMJ Quality & Safety, 22*(4), 273–277.

Siegrist, J. (2017). The effort–reward imbalance model. In C. L. Cooper & J. C. Quick (Eds.), *The Handbook of Stress and Health* (pp. 24–35). John Wiley & Sons

Siemieniuch, C. E., & Sinclair, M. A. (2006). Systems integration. *Applied Ergonomics, 37*(1), 91–110.

Siemieniuch, C. E., & Sinclair, M. A. (2014). Extending systems ergonomics thinking to accommodate the socio-technical issues of systems of systems. *Applied Ergonomics, 45*(1), 85–98.

Siemieniuch, C. E., & Sinclair, M. A. (2015). Methods in systems ergonomics. In J. R. Wilson & S. Sharples (Eds.), *Evaluation of Human Work* (Fourth ed., pp. 855–879). CRC Press

Simonse, L., Albayrak, A., & Starre, S. (2019). Patient journey method for integrated service design. *Design for Health, 3*(1), 82–97.

Sinclair, M. A. (2007). Ergonomics issues in future systems. *Ergonomics, 50*(12), 1957–1986.

Singleton, W. T. (1967). Ergonomics in systems design. *Ergonomics, 10*(5), 541–548.

Sinsky, C., & Linzer, M. (2020). Practice and policy reset post-COVID-19: Reversion, transition, or transformation? *Health Affairs, 39*(8), 1405–1411.

Smith, J. H., Cohen, W. J., Conway, F. T., Carayon, P., Derjani Bayeh, A., & Smith, M. J. (2002). Community ergonomics. In H. W. Hendrick & B. M. Kleiner (Eds.), *Macroergonomics: Theory, Methods, and Applications* (pp. 289–309). CRC Press.

Smith, J. H., & Smith, M. J. (1994). Community ergonomics: An emerging theory and engineering practice. In The Human Factors and Ergonomics Society (Ed.), *Proceedings of the Human Factors and Ergonomics Society 38th Annual Meeting* (pp. 729–733). The Human Factors and Ergonomics Society.

Smith, M. J., Carayon, P., Smith, J., Cohen, W., & Upton, J. (1994). Community ergonomics: A theoretical model for rebuilding the inner city. In The Human Factors and Ergonomics Society (Ed.), *The Human Factors and Ergonomics Society 38th Annual Meeting* (pp.724–728). The Human Factors and Ergonomics Society.

Smith, M. J., & Carayon-Sainfort, P. (1989). A balance theory of job design for stress reduction. *International Journal of Industrial Ergonomics*, *4*(1), 67–79.

Smith, M. J., Derjani-Bayeh, A., & Carayon, P. (2009). Community ergonomics and globalization: A conceptual model of social awareness. In C. M. Schlick (Ed.), *Industrial Engineering and Ergonomics* (pp. 57–66). Springer.

Soman, D. A., Koscelny, S. N., Neyens, D., Dietrich, A., Narasimhan, M., Taaffe, K., Allison, D., & Joseph, A. (2025). Using patient journey mapping and provider workflows to understand process barriers to pediatric mental and behavioral health care in emergency departments. *Applied Ergonomics*, *126*, 104512.

Stanton, N., Hedge, A., Brookhuis, K., Salas, E., & Hendrick, H. W. (Eds.). (2004). *Handbook of Human Factors and Ergonomics Methods*. CRC Press.

Steege, L. M., & Dykstra, J. G. (2016). A macroergonomic perspective on fatigue and coping in the hospital nurse work system. *Applied Ergonomics*, *54*, 19–26.

Steele, M. L., Talley, B., & Frith, K. H. (2018). Application of the SEIPS model to analyze medication safety in a crisis residential center. *Archives of Psychiatric Nursing*, *32*(1), 7–11.

Stiggelbout, A. M., Pieterse, A. H., & De Haes, J. C. (2015). Shared decision making: Concepts, evidence, and practice. *Patient Education and Counseling*, *98*(10), 1172–1179.

Still, C., Hoel, S., Strayer, A., Werner, N. E., & Fields, B. (2025). No place to go: The relationship between care partner inclusion practices and the hospital environment. *HERD*, *18*(1), 108–121.

Stone, P. W., Clarke, S. P., Cimiotti, J., & Correa-de-Araujo, R. (2004). Nurses' working conditions: Implications for infectious disease. *Emerging Infectious Diseases*, *10*(11), 1984–1989.

Sujan, M. (2024). Safety-II. In C. Cox, H. Hughes, & J. Nicholls (Eds.), *Patient Safety - Emerging Applications of Safety Science* (pp. 65–78). Class Professional Publishing.

Sujan, M., Crumpton, E., Finch, V., & Combes, J. (2024). From vision to reality: How human factors can inform the design of cardiac diagnostic services in community settings. *Human Factors in Healthcare*, *5*, 100075.

Sujan, M., Lounsbury, O., Pickup, L., Kaya, G. K., Earl, L., & McCulloch, P. (2024). What kinds of insights do Safety-I and Safety-II approaches provide? A critical reflection on the use of SHERPA and FRAM in healthcare. *Safety Science*, *173*, 106450.

Sujan, M., Pickup, L., Vosper, H., & Catchpole, K. (2021). *Making Human Factors and Ergonomics Work in Health and Social Care*. Chartered Institute of Ergonomics & Human Factors.

Sujan, M., Pickup, L., Vosper, H., & Catchpole, K. (2025). *Human Factors and Ergonomics in Health and Social Care - An Applied Approach*. Class Professional Publishing.

Sujan, M., Pool, R., & Salmon, P. (2022). Eight human factors and ergonomics principles for healthcare artificial intelligence. *BMJ Health & Care Informatics*, *29*(1), e100516.

Suri, J. F. (2001). Future challenges and opportunities for empathy in our science. *Ergonomics*, *44*(14), 1278–1289.

Sutcliffe, J., & Lomax, S. (2025). Lessons learned from implementing a Human Factors programme into an NHS trust. *Ergonomics*, *68*(5), 683–696.

Sutherland, A. B., Phipps, D. L., Grant, S., Hughes, J., Tomlin, S., & Ashcroft, D. M. (2024). Understanding the informal aspects of medication processes to maintain patient safety in hospitals: A sociotechnical ethnographic study in paediatric units. *Ergonomics*, *68*(3), 444–458.

Sznelwar, L. I., Silva, M. T., & Mascia, F. L. (2008). Working in public health services in Brazil: The relationship between different rationalities. *Applied Ergonomics*, *39*(4), 500–508.

Tarnow-Mordi, W. O., Hau, C., Warden, A., & Shearer, A. J. (2000). Hospital mortality in relation to staff workload: A 4-year study in an adult intensive care unit. *Lancet*, *356*, 185–189.

Tarver, W. L., Savoy, A., Patel, H., Weiner, M., & Holden, R. J. (2024). Inefficient processes and associated factors in primary care nursing: System configuration analysis. *JMIR Human Factors*, *11*, e49691.

Tawfik, D. S., Scheid, A., Profit, J., Shanafelt, T., Trockel, M., Adair, K. C., Sexton, J. B., & Ioannidis, J. P. A. (2019). Evidence relating health care provider burnout and quality of care: A systematic review and meta-analysis. *Annals of Internal Medicine*, *171*(8), 555–567.

Taylor, C., Maben, J., Jagosh, J., Carrieri, D., Briscoe, S., Klepacz, N., & Mattick, K. (2024). Care Under Pressure 2: A realist synthesis of causes and interventions to mitigate psychological ill health in nurses, midwives and paramedics. *BMJ Quality & Safety*, *33*(8), 523–538.

Taylor, J. A., Dominici, F., Agnew, J., Gerwin, D., Morlock, L., & Miller, M. R. (2012). Do nurse and patient injuries share common antecedents? An analysis of associations with safety climate and working conditions. *BMJ quality & safety*, *21*(2), 101–111.

Taylor-Adams, S., & Vincent, C. (2004). Systems analysis of clinical incidents: The London protocol. *Clinical Risk*, *10*(6), 211–220.

Tetteh, E., Wang, T., Kim, J. Y., Smith, T., Norasi, H., Van Straaten, M. G., Lal, G., Chrouser, K. L., Shao, J. M., & Hallbeck, M. S. (2024). Optimizing ergonomics during open, laparoscopic, and robotic-assisted surgery: A review of surgical ergonomics literature and development of educational illustrations. *American Journal of Surgery*, *235*, 115551.

Thatcher, A., & Yeow, P. H. P. (2016). A sustainable system of systems approach: A new HFE paradigm. *Ergonomics*, *59*(2), 167–178.

The Joint Commission. (2012). *Improving Patient and Worker Safety: Opportunities for Synergy, Collaboration and Innovation*. The Joint Commission.

Thomas, E. J., Sexton, J. B., Neilands, T. B., Frankel, A., & Helmreich, R. L. (2005). The effect of executive walk rounds on nurse safety climate attitudes: A randomized trial of clinical units. *BMC Health Services Research*, *5*(1), 28.

Tischendorf, J., Brunner, M., Knobloch, M. J., Schulz, L., Barker, A., Wright, M. O., Lepak, A., & Safdar, N. (2020). Evaluation of a successful fluoroquinolone restriction intervention among high-risk patients: A mixed-methods study. *PloS one*, *15*(8), e0237987.

Tremoulet, P., Krishnan, R., Karavite, D., Muthu, N., Regli, S. H., Will, A., & Michel, J. (2018). A heuristic evaluation to assess use of after visit summaries for supporting continuity of care. *Applied Clinical Informatics*, *9*(03), 714–724.

Trinkoff, A. M., Johantgen, M., Storr, C. L., Gurses, A. P., Liang, Y., & Han, K. (2011a). Linking nursing work environment and patient outcomes. *Journal of Nursing Regulation*, *2*(1), 10–16.

Trinkoff, A. M., Johantgen, M., Storr, C. L., Gurses, A. P., Liang, Y., & Han, K. (2011b). Nurses' work schedule characteristics, nurse staffing, and patient mortality. *Nursing Research*, *60*(1), 1–8.

Trist, E. (1981). *The Evolution of Socio-technical Systems*. Quality of Working Life Center.

Trist, E. L., & Bamforth, K. (1951). Some social and psychological consequences of the longwall method of coal getting. *Human Relations*, *4*, 3–39.

Tucker, A. L., & Edmondson, A. C. (2003). Why hospitals don't learn from failures: Organizational and psychological dynamics that inhibit system change. *California Management Review*, *45*(2), 55–72.

Tucker, A. L., Edmondson, A. C., & Spear, S. (2002). When problem solving prevents organizational learning. *Journal of Organizational Change Management*, *15*(2), 122–137.

Tucker, A. L., & Spear, S. J. (2006). Operational failures and interruptions in hospital nursing. *Health Services Research*, *41*(3 Part 1), 643–662.

Ulrich, R. S., Zimring, C., Barch, X. Z., Dubose, J., Seo, H. B., Choi, Y. S., Quan, X., & Joseph, A. (2008). A review of the research literature on evidence-based healthcare design. *HERD*, *1*(3), 61–125.

Valdez, R. S., & Holden, R. J. (Eds.). (2021). *The Patient Factor: Applications of Patient Ergonomics*. CRC Press.

Valdez, R. S., Holden, R. J., Novak, L. L., & Veinot, T. C. (2015). Transforming consumer health informatics through a patient work framework: Connecting patients to context. *Journal of the American Medical Informatics Association*, *22*(1), 2–10.

Valdez, R. S., Lyon, S. E., Corbett, J. P., Wellbeloved-Stone, C., Hasan, S., Taylor, L., DeBoer, M. D., Cherñavvsky, D., & Patek, S. D. (2024). Macroergonomic components of the patient work system shaping dyadic care management during adolescence: A case study of type 1 diabetes. *Ergonomics*, *67*(11), 1575–1595.

van Boeijen, A., Daalhuizen, J., Zijlstra, J. G., & van der Schoor, R. (2013). *Delft Design Guide*. BIS Publishers.

Vartiainen, M., & Ruohomaki, V. (1994). Simulation games as tools for work development, and their psychological bases. In G. Bradley & H. W. Hendrick (Eds.), *Human Factors in Organizational Design and Management-IV* (pp. 73–78). Elsevier.

Verhagen, M. J., de Vos, M. S., Sujan, M., & Hamming, J. F. (2022). The problem with making Safety-II work in healthcare. *BMJ Quality & Safety*, *31*, 402–408.

Vicente, K. J. (1999). *Cognitive Work Analysis*. Lawrence Erlbaum Associates.

Vicente, K. J., Burns, C. M., & Pawlak, W. S. (1998). Better handbooks, better design. *Ergonomics In Design*, *6*(2), 21–27.

Vincent, C. (2010). *Patient Safety*. Wiley-Blackwell.

Vincent, C., & Amalberti, R. (2016). *Safer Healthcare - Strategies for the Real World*. Springer Open.

Vincent, C., Irving, D., Bellandi, T., Higham, H., Michel, P., Staines, A., Adams, S., Brown, J., Hibbert, P. D., Hemmelgarn, C., Joseph, L., Pires, K., Sheridan, S., Sunol, R., Ushiro, S., Wu, A. W., & Zambon, L. (2025). Systems analysis of clinical incidents: Development of a new edition of the London Protocol. *BMJ Quality & Safety*, *34*(6), 413–420.

Vosper, H., Hignett, S., & Bowie, P. (2018). Twelve tips for embedding human factors and ergonomics principles in healthcare education. *Medical Teacher*, *40*, 357–363.

Wachter, R. M. (2010). Patient safety at ten: Unmistakable progress, troubling gaps. *Health Affairs*, *29*(1), 165–173.

Wåhlin, C., Kvarnström, S., Öhrn, A., & Nilsing Strid, E. (2019). Patient and healthcare worker safety risks and injuries. Learning from incident reporting. *European Journal of Physiotherapy*, *22*(1), 44–50.

Walker, G. H., Stanton, N. A., Salmon, P. M., Jenkins, D. P., & Rafferty, L. (2010). Translating concepts of complexity to the field of ergonomics. *Ergonomics*, *53*(10), 1175–1186.

Walker, J., & Carayon, P. (2009). From tasks to processes: The case for changing health information technology to improve health care. *Health Affairs*, *28*(2), 467–477.

Walsh, K. E., Bundy, D. G., & Landrigan, C. P. (2014). Preventing health care–associated harm in children. *JAMA*, *311*(17), 1731–1732.

Waters, T. R. (2010). Introduction to ergonomics for healthcare workers. *Rehabilitation Nursing*, *35*(5), 185–191.

Waterson, P. (2009). A critical review of the systems approach within patient safety research. *Ergonomics*, *52*(10), 1185–1195.

Waterson, P. (2023). Ways of seeing (and not seeing) safety. In J.-C. Le Coze & T. Reiman (Eds.), *Visualizing Safety, An Exploration - Drawings, Pictures, Images, Videos and Movies* (pp. 35–40). Springer.

Waterson, P., Robertson, M. M., Cooke, N. J., Militello, L., Roth, E., & Stanton, N. A. (2015). Defining the methodological challenges and opportunities for an effective science of sociotechnical systems and safety. *Ergonomics, 58*(4), 565–599.

Watson, M., Sanderson, P., & Russell, W. J. (2004). Tailoring reveals information requirements: The case of anaesthesia alarms. *Interacting with Computers, 16*(2), 271–293.

Watt, A., Jun, G. T., & Waterson, P. (2019). Resilience in the blood transfusion process: Everyday and long-term adaptations to 'normal' work. *Safety Science, 120*, 498–506.

Wears, R. L., Perry, S. J., & Patterson, E. S. (2012). Handoffs and transitions of care. In P. Carayon (Ed.), *Handbook of Human Factors and Ergonomics in Health Care and Patient Safety* (Second ed., pp. 163–172). CRC Press.

Wears, R. L., Perry, S. J., Wilson, S., Galliers, J., & Fone, J. (2006). Emergency department status boards: User-evolved artefacts for inter- and intra-group coordination. *Cognition, Technology & Work, 9*(3), 163–170.

Wears, R. L., & Sutcliffe, K. M. (2020). *Still Not Safe - Patient Safety and the Middle-Managing of American Medicine*. Oxford University Press.

Weaver, S., Stewart, K., & Kay, L. (2021). Systems-based investigation of patient safety incidents. *Future Healthcare Journal, 8*(3), e593–e597.

Weick, K. E., & Quinn, R. E. (1999). Organizational change and development. *Annual Review of Psychology, 50*, 361–386.

Weick, K. E., & Sutcliffe, K. M. (2001). *Managing the Unexpected: Assuring High Performance in an Age of Complexity*. Jossey-Bass.

Weiler, D. T., Lingg, A. J., Eagan, B. R., Shaffer, D. W., & Werner, N. E. (2022). Quantifying the qualitative: Exploring epistemic network analysis as a method to study work system interactions. *Ergonomics, 65*(10), 1434–1449.

Weinger, M. B., Wiklund, M., & Gardner-Bonneau, D. J. (Eds.). (2011). *Handbook of Human Factors in Medical Device Design*. Taylor & Francis Group.

Werner, N. E. (2025). *It's time to design the caregiver job*. unpublished work.

Werner, N. E., Malkana, S., Gurses, A. P., Leff, B., & Arbaje, A. I. (2017). Toward a process-level view of distributed healthcare tasks: Medication management as a case study. *Applied Ergonomics, 65*, 255–268.

Werner, N. E., Ponnala, S., Doutcheva, N., & Holden, R. J. (2020). Human factors/ergonomics work system analysis of patient work: State of the science and future directions. *International Journal for Quality in Health Care, 33*(Supplement_1), 60–71.

Werner, N. E., Rutkowski, R., Graske, A., Finta, M. K., Sellers, C. R., Seshadri, S., & Shah, M. N. (2020). Exploring SEIPS 2.0 as a model for analyzing care transitions across work systems. *Applied Ergonomics, 88*, 103141.

Werner, N. E., Rutkowski, R. A., Holden, R. J., Ponnala, S., & Gilmore-Bykovskyi, A. (2022). A human factors and ergonomics approach to conceptualizing care work among caregivers of people with dementia. *Applied Ergonomics, 104*, 103820.

Werner, N. E., Rutkowski, R. A., Krause, S., Barton, H. J., Wust, K., Hoonakker, P., King, B., Shah, M. N., Pulia, M. S., Brenny-Fitzpatrick, M., Smith, M., & Carayon, P. (2021). Disparate perspectives: Exploring healthcare professionals' misaligned mental models of older adults' transitions of care between the emergency department and skilled nursing facility. *Applied Ergonomics, 96*, 103509.

Werner, N. E., Smith, M., Rutkowski, R. A., Barton, H. J., Wust, K., Hoonakker, P., King, B. J., Shah, M. N., Patterson, B. W., Pulia, M. S., Dail, P. V. W., & Carayon, P. (2025). Engineering safe care journey: Designing a patient safety passport. *Journal of Patient Safety*.

Werner, N. E., Tong, M., Borkenhagen, A., & Holden, R. J. (2019). Performance-shaping factors affecting older adults' hospital-to-home transition success: A systems approach. *The Gerontologist, 59*(2), 303–314.

Wetterneck, T. B., Skibinski, K. A., Roberts, T. L., Kleppin, S. M., Schroeder, M., Enloe, M., Rough, S. S., Hundt, A. S., & Carayon, P. (2006). Using failure mode and effects analysis to plan implementation of Smart intravenous pump technology. *American Journal of Health-System Pharmacy*, *63*, 1528–1538.

Wetterneck, T. B., Walker, J. M., Blosky, M. A., Cartmill, R. S., Hoonakker, P., Johnson, M. A., Norfolk, E., & Carayon, P. (2011). Factors contributing to an increase in duplicate medication order errors after CPOE implementation. *Journal of the American Medical Informatics Association*, *18*(6), 774–782.

Wheway, J. L., & Jun, G. T. (2021). Adopting systems models for multiple incident analysis: Utility and usability. *International Journal for Quality in Health Care*, *33*(4), 1–8.

Wiig, S., Calderwood, C. J., & O'Hara, J. (2024). Sailing too close to the wind? How harnessing patient voice can identify drift towards boundaries of acceptable performance. *Healthcare*, *12*(15), 1532.

Williams, E. S., Halbesleben, J. R. B., Manwell, L. B., McMurray, J. E., Rabatin, J., Rashid, A., & Linzer, M. (2012). The effect of workplace health care worker stress and burnout on patient outcomes. In P. Carayon (Ed.), *Handbook of Human Factors and Ergonomics in Health Care and Patient Safety* (Second ed., pp. 119–131). Taylor & Francis Group.

Willis, C. D., Saul, J., Bevan, H., Scheirer, M. A., Best, A., Greenhalgh, T., Mannion, R., Cornelissen, E., Howland, D., Jenkins, E., & Bitz, J. (2016). Sustaining organizational culture change in health systems. *Journal of Health Organization and Management*, *30*(1), 2–30.

Wilson, J. R. (1995). Ergonomics and participation. In J. R. Wilson & E. N. Corlett (Eds.), *Evaluation of Human Work* (Second ed., pp. 1071–1096). Taylor & Francis Group.

Wilson, J. R. (2000). Fundamentals of ergonomics in theory and practice. *Applied Ergonomics*, *31*(6), 557–567.

Wilson, J. R. (2014). Fundamentals of systems ergonomics/human factors. *Applied Ergonomics*, *45*(1), 5–13.

Wilson, J. R., & Corlett, N. (Eds.). (2005). *Evaluation of Human Work* (Third ed.). CRC Press.

Wilson, J. R., & Sharples, S. (2015). Methods in the understanding of human factors. In J. R. Wilson & S. Sharples (Eds.), *Evaluation of Human Work* (Fourth ed., pp. 1–32). CRC Press.

Wilson, K. A., Burke, C. S., Priest, H. A., & Salas, E. (2005). Promoting health care safety through training high reliability teams. *Quality & Safety in Health Care*, *14*(4), 303–309.

Wilson, P., & Kislov, R. (2022). *Implementation Science*. Cambridge University Press.

Wood, L. J., & Wiegmann, D. A. (2020). Beyond the corrective action hierarchy: A systems approach to organizational change. *International Journal for Quality in Health Care*, *32*(7), 438–444.

Woods, D. (2000). *Broad-Based Systems Approaches*. Quality Interagency Coordination Task Force (QuIC).

Woods, D. D. (2006). Essential characteristics of resilience. In E. Hollnagel, D. D. Woods, & N. Leveson (Eds.), *Resilience Engineering - Concepts and Precepts* (pp. 21–34). Ashgate.

Woodward, S. (2019). Moving towards a safety II approach. *Journal of Patient Safety and Risk Management*, *24*(3), 96–99.

Wooldridge, A. R., Carayon, P., Hoonakker, P., Hose, B.-T., Ross, J., Kohler, J. E., Brazelton, T., Eithun, B., Kelly, M. M., Shannon, M. D., Rusy, D., Durojaiye, A., & Gurses, A. P. (2019). Complexity of the pediatric trauma care process: Implications for multi-level awareness. *Cognition, Technology & Work*, *21*, 397–416.

Wooldridge, A. R., Carayon, P., Hoonakker, P., Hose, B. Z., Eithun, B., Brazelton, T., Ross, J., Kohler, J. E., Kelly, M. M., Dean, S. M., Rusy, D., & Gurses, A. P. (2020). Work system barriers and facilitators in inpatient care transitions of pediatric trauma patients. *Applied Ergonomics*, *85*, 103059.

Wooldridge, A. R., Carayon, P., Hoonakker, P., Hose, B.-Z., Schroeer, K., Brazelton, T., Eithun, B., Rusy, D., Ross, J., Kohler, J., Kelly, M. M., Dean, S., Springman, S., Rahal, R., & Gurses, A. P. (2022). Care transition of trauma patients: Processes with articulation work before and after handoff. *Applied Ergonomics, 98*, 103606.

Wooldridge, A. R., Carayon, P., Hundt, A. S., & Hoonakker, P. L. T. (2017). SEIPS-based process modeling in primary care. *Applied Ergonomics, 60*, 240–254.

Wooldridge, A. R., Morgan, J., Ramadhani, W. A., Hanson, K., Vazquez-Melendez, E., Kendhari, H., Shaikh, N., Riech, T., Mischler, M., Krzyzaniak, S., Barton, G., Formella, K. T., Abbott, Z. R., Farmer, J. N., Ebert-Allen, R., & Croland, T. (2024). Interactions in sociotechnical systems: Achieving balance in the use of an augmented reality mobile application. *Human Factors, 66*(3), 658–682.

World Health Organization (WHO). (2009). *Conceptual Framework for the International Classification for Patient Safety*. WHO Press.

World Health Organization (WHO). (2010). *Healthy Workplaces: A Model for Action*. WHO Press.

World Health Organization (WHO). (2019a). *Burn-out an "occupational phenomenon": International Classification of Diseases*. WHO Press.

World Health Organization (WHO). (2019b). *Medication Safety in Transitions of Care*. WHO Press.

World Health Organization (WHO). (2021). *Global Patient Safety Action Plan 2021–2030*. WHO Press.

World Health Organization (WHO). (2024). *Global Patient Safety Report 2024*. WHO Press.

Wrzesniewski, A., & Dutton, J. E. (2001). Crafting a job: Revisioning employees as active crafters of their work. *The Academy of Management Review, 26*(2), 179–201.

Wu, A. W., Sax, H., Letaief, M., Bellandi, T., Newman-Toker, D., Paine, L. A., Vincent, C., Pronovost, P. J., Elorrio, E. G., Kachalia, A., & Haut, E. R. (2020). COVID-19: The dark side and the sunny side for patient safety. *Journal of Patient Safety and Risk Management, 25*(4), 137–141.

Wulff, I. A., Westgaard, R. H., & Rasmussen, B. (1999a). Ergonomic criteria in large-scale engineering design - I - Management by documentation only? Formal organization vs. designers' perceptions. *Applied Ergonomics, 30*(3), 191–205.

Wulff, I. A., Westgaard, R. H., & Rasmussen, B. (1999b). Ergonomic criteria in large-scale engineering design - II - Evaluating and applying requirements in the real-world of design. *Applied Ergonomics, 30*(3), 207–221.

Wust, K., Barton, H., Werner, N. E., Rutkowski, R., Hoonakker, P. L. T., Shah, M. N., Patterson, B. W., Pulia, M. S., Buckley, D., Smith, M., King, B., Dail, P. V., & Carayon, P. (2022). SEIPS-based patient journey mapping: Application to a patient visit in the emergency department. In The Human Factors and Ergonomics Society (Ed.), *Proceedings of the Human Factors and Ergonomics Society Annual Meeting* (Vol. 66, pp. 848–852). The Human Factors and Ergonomics Society.

Wust, K., Barton, H. J., Werner, N., Hoonakker, P., Phelps, K., Dail, P. V. W., & Carayon, P. (2023). A community-based usability evaluation of emergency department discharge instructions. *14th Organizational Design and Management Conference*, Bordeaux, France.

Wust, K. L., Carayon, P., Werner, N. E., Hoonakker, P. L. T., Barton, H., Dail, P. V. W., King, B., Patterson, B. W., Shah, M. S., & Buckley, D. (2025). Work system barriers and facilitators to collaboration during the Emergency Department discharge process: Comparison of multiple perspectives. *Submitted.*

Xie, A., & Carayon, P. (2015). A systematic review of human factors and ergonomics (HFE)-based healthcare system redesign for quality of care and patient safety. *Ergonomics, 58*(1), 33–49.

Xie, A., Carayon, P., Cartmill, R., Li, Y., Cox, E. D., Plotkin, J. A., & Kelly, M. M. (2015). Multi-stakeholder collaboration in the redesign of family-centered rounds process. *Applied Ergonomics*, *46*(Part A), 115–123.

Xie, A., Carayon, P., Cox, E. D., Cartmill, R., Li, Y., Wetterneck, T. B., & Kelly, M. M. (2015). Application of participatory ergonomics to the redesign of the family-centred rounds process. *Ergonomics*, *58*(10), 1726–1744.

Xie, A., Carayon, P., Kelly, M. M., Li, Y., Cartmill, R., DuBenske, L. L., Brown, R. L., & Cox, E. D. (2012). Managing different perspectives in the redesign of family-centered rounds in a pediatric hospital. In The Human Factors and Ergonomics Society (Ed.), *Proceedings of the Human Factors and Ergonomics Society 56th Annual Meeting* (pp. 1733–1737). The Human Factors and Ergonomics Society.

Xie, A., Rock, C., Hsu, Y. J., Osei, P., Andonian, J., Scheeler, V., Keller, S. C., Cosgrove, S. E., & Gurses, A. P. (2018). Improving daily patient room cleaning: An observational study using a human factors and systems engineering approach. *IISE Transaction on Occupational Ergonomics and Human Factors*, *6*(3–4), 178–191.

Xie, A., Sax, H., Daodu, O., Alam, L., Sultan, M., Rock, C., Stewart, C. M., Perry, S. J., & Gurses, A. P. (2024). Environmental cleaning and disinfection in the operating room: A systematic scoping review through a human factors and systems engineering lens. *Infection Control & Hospital Epidemiology*, *45*(7), 880–889.

Yanke, E., Carayon, P., & Safdar, N. (2014). Commentary: Translating evidence into practice using a systems engineering framework for infection prevention. *Infection Control and Hospital Epidemiology*, *35*(9), 1176–1182.

Yanke, E., Moriarty, H., Carayon, P., & Safdar, N. (2018). A qualitative, interprofessional analysis of barriers to and facilitators of implementation of the Department of Veterans Affairs' Clostridium difficile prevention bundle using a human factors engineering approach. *American Journal of Infection Control*, *46*(3), 276–284.

Yardley, S., Williams, H., Bowie, P., Edwards, A., Noble, S., Donaldson, L., & Carson-Stevens, A. (2022). Which human factors design issues are influencing system performance in out-of-hours community palliative care? Integration of realist approaches with an established systems analysis framework to develop mid-range programme theory. *BMJ Open*, *12*(1), e048045.

Yoo, S., Lee, K. H., Baek, H., Ryu, B., Chung, E., Kim, K., Yi, J. C., Park, S. B., & Hwang, H. (2015). Development and user research of a smart bedside station system toward patient-centered healthcare system. *Journal of Medical Systems*, *39*(9), 1–11.

Yu, A., Chopra, V., Mueller, S. K., Wray, C. M., & Jones, C. D. (2024). Engineering safe care journeys: Reenvisioning interhospital transfers. *Journal of Hospital Medicine*, *19*(7), 629–634.

Zhang, J., Johnson, T. R., Patel, V. L., Paige, D. L., & Kubose, T. (2003). Using usability heuristics to evaluate patient safety of medical devices. *Journal of Biomedical Informatics*, *36*, 23–30.

Zhang, S., & Gu, X. (2025). Utilising human factors systems approach to support the well-being of healthcare workers: A qualitative study of work system interactions during the COVID-19 pandemic. *Cognition, Technology & Work*, 1–19.

Zimring, C., Jacob, J. T., Denham, M. E., Kamerow, D. B., Hall, K. K., Cowan, D. Z., Kasali, A., Lenfestey, N., Do, E., & Steinberg, J. P. (2013). The role of facility design in preventing the transmission of healthcare-associated infections: Background and conceptual framework. *HERD*, *7*(Suppl), 18–30.

Zink, K. (2000). Ergonomics in the past and the future: From a German perspective to an international one. *Ergonomics*, *43*(7), 920–930.

Zink, K. J. (2006). Human factors, management and society. *Theoretical Issues in Ergonomics Science*, *7*(4), 437–445.

Zisberg, A., Agmon, M., Gur-Yaish, N., Rand, D., Hayat, Y., Gil, E., & WALK-FOR team. (2018). No one size fits all-the development of a theory-driven intervention to increase in-hospital mobility: The "WALK-FOR" study. *BMC Geriatrics*, *18*(1), 91.

Index

Pages in *italics* refer to figures and pages in **bold** refer to tables.

E

F

G

H

I

J

K

L

M

N

O

P

Q

R

S

T

U

V

W